# Secrets
## *of the* Mind

# A.G. Cairns-Smith

# Secrets
## *of the* Mind

## A Tale of Discovery
## *and* Mistaken Identity

C

COPERNICUS
AN IMPRINT OF SPRINGER-VERLAG

Published in the United States by Copernicus, an imprint of Springer-Verlag New York, Inc.

　　Copernicus
　　Springer-Verlag New York, Inc.
　　175 Fifth Avenue
　　New York, NY 10010


Library of Congress Cataloging-in-Publication Data
Cairns-Smith, A.G. (Alexander Graham)
　　Secrets of the mind : a tale of discovery and mistaken identity /
　A.G. Cairns-Smith.
　　　　p.　cm.
　　Includes bibliographical references (p.　　) and index.
　　ISBN 0-387-98692-8 (alk. paper)
　　1. Consciousness.　2. Perception.　I. Title.
　BF311.C147　1999
　153—dc21　　　　　　　　　　　　　　　　　99-14663


Manufactured in the United States of America.
Printed on acid-free paper.

9 8 7 6 5 4 3 2 1

ISBN 0-387-98692-8　　SPIN 10706577

*To Dorothy Anne*

# Preface

We go to a performance of Shakespeare's *Othello* and are moved by it, amazed yet convinced of the powerful, disastrous effects of envy, jealousy and lust. On the way home we feel like stopping off for an ice cream and then decide to have coffee as well. We may not be operating on the same plane of intensity as the great Othello or the vile Iago, but we too have been following our feelings. We do it all the time. It is part of the point of being conscious. It helps us to understand the behaviour of those around us. Yet current science, in its hugely successful exploration of Nature, has no satisfactory way of including feelings and emotions in its explanations of behaviour. Nature seems to be holding something back. What is her secret? And where? We will be embarking shortly on a modest voyage of discovery.

We will find that there are several secrets of the mind, and that they are of different kinds with clues to be found in different places. First of all, there are the open secrets. These are things we all know because they are within us. We know that we have feelings and sensations, emotions and moods; that these may come as nice or nasty or neutral; that they may be anything from barely perceptible to overwhelming. From our inner experience of their qualities we just know how pain and hunger, fear and lust *work*. Such elements of our consciousness nudge us or push us to behave in ways that are, on the whole, appropriate to our own survival and the continued propagation of our kind. They are eminently understandable in these terms and in the terms of life and literature, but not yet in terms of physics and chemistry; not at all in terms of the molecules and forces through which we can describe so well how genes and brain cells work.

In darker rocky places we come across an ominous secret of Nature that is revealed by today's science but whose implications have yet to be taken seriously. Feelings and sensations cannot be part of some other world, some "world of the mind" immune to the methods of science as some would have us believe. They must be fully part of the physical world or the means to produce them could not have evolved. This may sound like an innocent enough thing to say, and most scientists would probably let it through on the nod. Yet I think that it is set to unhinge the whole view of the world that science currently presents.

All of this is not to deny that a now-thriving brain science is discovering navigable and hitherto uncharted waters where new secrets are being uncovered almost daily about underlying mechanisms of perception, thought and action. Nor is it to deny that the Secret Agent of our brain-made conscious mind is indeed some kind of device on a par with other parts of us. In engineering terms it is an auxiliary control device. But if we are trying to think as engineers about its workings, I think we will be defeated; not so much because it is beyond our intellectual capacity, but because our science is not, as yet, up to it.

*Secrets of the Mind* makes its journey within the vast and complicated seas of the human mind. In this respect it is like its predecessor: *Evolving the Mind*. But it makes a shorter journey requiring, I suppose, less skills of scientific seamanship. And yet it is not just a tourist's version. It is a different voyage. It starts from a different place. It starts in mind, not matter. *Secrets* starts, and finishes, with the question of whether or not we have free will. It was inspired by that age-old question, as perhaps sailors may once have been inspired to travel in search of the ends of the Earth. And if, as it seems likely, the age-old question will no more be satisfied than was the sailors' age-old quest, it is of some interest to try to see why.

I have put familiar or unfamiliar mini-quotes from the great psychologist Shakespeare under the chapter titles to remind us that it is not only in laboratories and studies that the secrets of the mind are to be unravelled.

I am grateful to many friends and colleagues who have helped in discussions and arguments and in providing, very often, wider opportunities for more. I started writing the book during a two-month stay at UCLA, in the Center for the Study of Evolution and the Origin of Life, at the kind invitation of Bill Schopf. While I was over there another old friend, Sherwood Chang, gave me the chance to present my ideas at NASA-Ames—to more bright minds—and *over here* Jack Cohen has been similarly catalytic. Then I have special thanks for those who were kind enough to read the beast, or bleeding chunks of it, in its earlier or later forms: Bruce Charlton, Kee Dewdney, Peter Marcer, David Mauzerall, Peter Munz, John Nixon, Pippa Orr, Frank Spence, my daughters Sarah and Emma, and especially for the detailed scrutiny of the final versions by Jonathan Cobb of Copernicus. And then, as always, I have most thanks of all for my wife Dorothy Anne.

Graham Cairns-Smith
Uplawmoor, January 1999

# Sources of Quotations

Quotations under chapter titles are from the plays of Shakespeare

# Contents

# *Preview*

We can do what we want. We do what we feel like doing much of the time. That's what most people think. But I was taught that Science denies this, that Science says that we are deluding ourselves if we think that what we *feel* can have anything directly to do with our behaviour; because things like that—feelings, sensations, pains, pleasures, and so on—are subjective and so cannot enter into any scientific account of anything. A scientific account has to be, above all, objective. I was taught that Science only deals with things that are out there in the real world for all to see; that although we may think that our feelings matter, this too is just a subjective impression. What we do, how we behave, *really* depends on cells and circuits in our brains whose actions are largely predictable, at least in principle. And if in practice we are not able to see exactly how everything is going to turn out, in brains or anywhere else, this is just because we cannot know enough about the details of what's going on in the world; or because there is some play, some randomness in Nature's machinery; or because the math is too difficult. Predictable or random, in neither can there be any of this "free will" that people talk of.

That is what I was taught. But is it true?

It has turned out that most of what the brain does in perception and thought, even in action, it does unconsciously; and this is the case even for what we call *conscious* perception, thought, or action. Most of that is also unconscious, what we can think of as computing—and much of it is understandable if not yet fully understood in decently objective terms. The Great Enigma is in the phenomena of consciousness itself.

The question that starts the book is the question of whether we have free will, but this is soon put aside for a much clearer issue: Do feelings as such contribute to our behaviour? Are they part of the brain's machinery over and above the neural machinery that we think we more or less understand? Or is it always possible to explain behaviour without bringing in the subjective world of our consciousness by staying with objective things like molecules, cells, circuits, and so on?

The very word "conscious" is, unhappily, a battleground—and a quagmire to be avoided as much as possible. I will just *say* what I mean by the word. I choose a restricted meaning so as to face rather than avoid that fascinating Enigma. It is not helpful to say that "conscious" is a synonym for "aware," as is commonly done; not if it is the essential nature of consciousness that we are trying to uncover, since science has now shown us just how much awareness is unconscious. The same goes for thought, which is not the essential thing about consciousness either: There is much unconscious thought. I will say that the essential features of consciousness are sensations and feelings—a sensation of pain, or of the colour yellow, or a feeling of anger or enthusiasm, and so on. Philosophers call these things *qualia*. I will say that it is specifically the associated sensations and feelings that make any piece of brain work con-scious. In our consciously controlled actions, for example, we set in train innumerable activities in the brain, in the spinal cord, in nerves and muscles, that are unconscious—automatic computer work. It is only if we go by the feel of it too that an action is a con-scious action. That is what I will say.

In Part II, I will draw your attention to a time bomb. The cen-tral idea of modern biology is the theory of evolution through natu-ral selection which makes it as clear as we could want that sensa-tions and feelings—qualia—must have both physical causes and physical effects: They must belong to the physical world. They could not have evolved their intricate appropriateness otherwise. This key point was made more than a hundred years ago. It is a dev-astating conclusion whose significance has yet to sink in: It is what I call "the bomb in the foundations of science." Today's science is incomplete since it lacks the means to connect qualia to activities, ultimately molecular activities, in the brain. It is a connection which must be there, but science lacks the language to make it. It will no longer do to say that feelings are "subjective" or "outside science." No, **qualia belong to the physical world** (write that on your T-shirt). And if we can't see how this could be, then we have some deeply wrong ideas about what this physical world is really like. Today's brain science is rubbing our noses in the incomplete-ness in our understanding.

But brain science is also pouring out more constructive infor-mation and insights of great relevance to our story and our quest. This provides meat for the five chapters of Part III. We start out by rummaging in the brain to see if we can find any of the machinery for making feelings and sensations, since we will have decided by now that there must be such machinery. We will be hunting for the qualia makers: "qualagens" as I will be calling our elusive quarry.

Elusive they certainly are. Yet we know of brain areas and circuitry that do participate in making feelings and sensations, and emotions and moods of different sorts. We know some of the chemical switches and adjusters of such things, partly from the effects of mind-bending drugs. Strokes and other forms of brain damage may visibly affect specific brain areas and disturb or destroy particular sensations. There is no question that qualia are brain made, and that the brain has special circuitry associated with their control. The contrivance is too evident for all this to be some accidental side effect of evolution. But how exactly qualia are made is still a mystery. Our preliminary conclusion at this stage will be that the concerted activities of myriads of tiny (molecular) structures create, or rather *are*, the feelings and sensations of our conscious being.

By chapter 10 we have launched into visual perception as the best-studied example of what goes on in the brain when we become aware of the world around us. In the first place, we are unconsciously aware of much of this. But even when we are consciously aware, when the picture of our world is being painted in qualia, in sensations of colour, and space, and movement, and so on, the production of qualia and their intricate arrangement is no doubt largely controlled by underlying unconscious processes. These processes are coming to be understood in ever more detail. What remains in the dark is just *how* the associated feelings and sensations are made at all, and how they are arranged and manipulated in the ever-changing world of our conscious being.

In Part IV, on "The Secret Agent," we start with the question of why we have conscious perceptions when unconscious ones can do so well (Often they do better, as any juggler will tell you.) Both conscious and unconscious perceptions are representations of the world, but neither gives us a direct view of the world as it is. Rather, they are representations of the world in other terms. Unconscious perceptions are representations in terms of the brain's computing activities, in terms of nerve fibre connections between brain cells, electrochemical signals being passed between them, and so on. Conscious perceptions are representations of the world in terms of qualia. In addition to its computing skills, the brain is a machine which makes and intricately arranges the various feelings and sensations of perception: sensations of colour, or space—or surprise perhaps. Yet what is the point? We cannot say that we see this perceptual representation and act accordingly. This leads to absurdity. We have to say instead that we *are*, among other things, this representation, or perceptual *model* as it is also called. It is a model that is both a representation of the world and part of our self.

That is what we have to say and it is crucial, but it still does not go far enough. There is no point in a model that has no effects. So it cannot be a passive model of the sort you put in a display cabinet and admire. It must be a working model. And this is where current research and ideas about "the body image" come in. Our much-spoken-of body image, it turns out, is not just an image, a kind of picture that we have of ourselves, but a very significant part of ourselves. To make any sense, the body image cannot be just a copycat image of our body. Yes, moving our body makes this image of it move in our mind. But if this image is actually to be a working model, it should operate the other way around too. It must be that when your body image moves its finger, then the corresponding real finger moves.

But what a thing to say!

Then we dig a few spades deeper on the question of those invisible qualagens. What are qualia such as feelings and sensations? What are they in physical terms? And what are the instruments in the brain—the qualagens—that produce them? Like so much else about living things, I see the qualagens as pieces of technology, clever pieces of engineering. Looked at from the point of view of an engineer, the conscious mind becomes then a *device*, a component of the body's control machinery. It is typical of devices in general (think of a light bulb, or a photocopier, or a radio . . .) that each depends critically on certain physical effects (current heating, electrostatic attraction, electromagnetic waves, or whatever). A device may seem like conjurer's magic to anyone who does not understand that it depends on the application of some critical effect in order to work. It will seem downright spooky if the effect itself is new. So this is the idea: The effect at the centre of this wonderful decision-making device that we call our conscious mind is itself new. It is not just another way of talking about some familiar effect such as an electrostatic force whose principles of action are well understood.

There are good reasons for saying this. The causal principles operating when, say, we react to pain or pleasure are new to physics—to put it mildly. Within physics (and chemistry and molecular biology and most of neurobiology for that matter) an event may take place because of an electrostatic field, a chance molecular encounter, a tendency for molecules to diffuse into wider territory, or for bigger things to move under the influence of gravity . . . and so on; but never to relieve pain, never for the fun of it, never in anticipation of a sensation, or to escape embarrassment. At least never do we need to use such terms or seriously think they have any relevance when we are talking physics, or chemistry, or computing. In describing behaviour, however, we use such terms all the time: They are essential to our understanding. So how are such different things as feelings and force fields to be seen belonging to the same world, as assuredly they must? We must look for common roots.

So here is the idea taken a step further. The underlying stuff of the world is energy, quantum energy from which, through atoms and molecules and higher-order structures, the fabrics of matter are woven. The device we seek weaves another kind of fabric out of quantum energy, an evanescent fabric of feeling and sensation and other such elements of consciousness.

So ends Part IV with a speculation on the wild side. But a deeper, more general, and I think much less speculative idea will have been the main one. It came to me as it has come to others "by osmosis"—as, say, a language may come to us more through unconscious than conscious effort. It had to be that way perhaps because the idea is so unpalatable that I hardly dare whisper it here. It is that we all belong, body and soul, to a physical world (although not quite the one described in current textbooks). We are objects in the Universe along with stars or stones or rivers or books or songs. It is not just our bodies that are objects in the physical world, but every part of us. There will be no holding back, say, the conscious bit of our mind. We are thoroughly *things* or we could not have evolved.

Think of some wonderful or beautiful thing. Think of the great waterfall at Iguacu, or the Isle of Rhum in the Scottish Hebrides, or the cathedral at Chartres, or an exquisite piece of jewelry . . . These are things. Should we be ashamed to be in such company? We too are *amazing* things.

The book ends with a brief recapitulation and (at last) a return, in the coda, to the opening topic of free will.

# *1*

# *Open Secrets*

## *from within*

*T*here are such important things about the mind which we all know, or think we know, but from which science still tends to avert her gaze. We know, or think we know, that feelings and sensations affect what we do: that they are, as it were, part of the mechanism of our behaviour. Shakespeare thought that. Most of us, most of the time, speak as if we do too. We know, or think we know, that there is a conscious part to us and an unconscious part to us, that they are distinct and that they both matter. But again, science tends to be a little embarrassed, hoping perhaps that it can stay with the language of electrochemistry, computing, and so on; hoping to explain the conscious part of us in terms that do not seriously confront such subjective things as feelings and sensations. Attitudes in science are changing here, but there is still much pussyfooting about. The brain itself is more straightforward: It makes feelings and sensations, no problem, as is becoming ever more evident.

# Chapter 1

# *Doing as we like*

*seems, madam! nay, it is; I know not "seems"*

*I* ask you to imagine that you have just been kidnapped by a group demanding independence for Outer Megalomania. Bound and gagged in the trunk of a stolen car, jammed between a bag of dirty clothes and a box of overripe vegetables, you have the distinct sense of having lost your freedom. You might decide to be philosophical about your situation and say to yourself, "Well, freedom is only relative after all. There are always limits of some sort. I mean, I can't just fly unaided, can I?—or even play the violin. *So* it's a bit more constrained in here than usual, but . . . ."

Or perhaps you are lucky enough to be a fatalist and believe that everything is predestined, that the future is all laid out pat. Then there would be nothing to worry about, at least nothing you could do about the situation. What will be, will be. The car swerves yet again. You think: "Perhaps the driver is fatalist. Should I pretend that I hold this view of the world so as to make friends with my kidnappers?"

I don't think I know any real card-carrying fatalists—I mean who really believe it and act as if they believe it. It seems to most people that we are usually more or less free to do as we like, to choose one course of action from a number of available possibilities; that we have, as they say, "free will." We seem to have even greater freedom when it comes to our thoughts. We can think as we like, can't we? We can make plans, choose from options we have made for ourselves. At least to some extent we can determine for ourselves what our future is going to be like.

That is what most people actually think, to judge from their behaviour and, it seems, with good reason. Seeing and choosing between possible courses of action? Why, this kind of thing is surely what minds are for.

Yet through the ages there have been fatalists of one brand or another saying that "everything is pre-determined," or something to that effect. In that case there could be no free will; only, at best, an illusion of it. Until well into the 20th century there was such a fatalism in science itself. It seemed inevitable: If the world consists of atoms moving according to rigid laws then it is all a kind of clockwork, which once set up and put into motion fixes the future for all time. But now the deepest theory of matter we have has changed the rules: According to the quantum theory, developed during the first three decades of the 20th century, nothing that happens on the atomic level is strictly predictable; it would be impossible even for an all-seeing being to foresee everything that will happen, even for a short time into the future.[1]

"Everything is the fruit of chance and necessity" said a philosopher in ancient Greece,[2] sounding quite liberal compared to fatalism. This way of thinking is close to that of our current science. But a chance-and-necessity doctrine, although not fatalistic, seems also to amount to a denial of free will.

The argument runs like this: Consider first a fatalistic clockwork world. In it, our actions are controlled by the inevitable meshing of "wheels within wheels" in our brains. They are automatic consequences of what we were at birth and what has happened to us since. In that case, all that we do, we do of *necessity*. There's no free will there, because there is no freedom. On the other hand if our decisions are made by *chance*, if there is some kind of lottery wheel among the clockwork, there is no free will there either, because there is no will.

Admittedly, a combination of chance and necessity is a good bit more interesting than either by itself. The greatest creative force we know of, evolution by natural selection, depends on such a combination. Life, as a phenomenon, can be understood pretty well in terms of chance and necessity.[3] And there is little doubt that the brain *does* have clockwork features as well as enough play in the mechanism to give plenty of scope for chance events to play a part too. Simple chance-and-necessity robots can be made that give the strong impression that they have wills of their own. No doubt a purely chance-and-necessity brain could give an onlooker the utterly convincing impression of wilfulness. If that great creator, evolution, needed only chance and necessity to make human brains, then perhaps that lesser creator, the brain itself, needs no more for its operations?

For sure this is a reasonable conjecture, but is it really true? Anyway, there seems to be more to it. The doubt arises from a secret we have all been let in on, but which is still alien to objective science. Whether or not we have what might be called free will, we all know, or think we know, that it is not only chance and clockwork that explains our behaviour: There are also feelings and sensations to nudge us, persuade us, drive us.

Feelings can drive us? Shakespeare certainly thought so. Here is a famous passage from *Othello*:

IAGO   O, beware, my lord, of jealousy;
        It is the green-eyed monster, which doth mock
        The meat it feeds on: that cuckold lives in bliss
        Who, certain of his fate, loves not his wronger;
        But, O, what damned minutes tells he o'er
        Who dotes, yet doubts, suspects, yet strongly loves![4]

Iago hates Othello for his success. He hates Othello because Othello did not promote him, because Othello is a Moor. But here Iago is playing friend and councilor to his master, acting as a kind of malignant psychiatrist—uncovering feelings, suggesting dire ideas, opening wounds. The whole of this play runs on feelings, and it would be incomprehensible if we did not directly know about them from our own experiences. This knowledge corresponds to knowledge of engineering principles needed to understand how, say, a clock or a bicycle works. Indeed, all of Shakespeare's plays attest to the role of feelings and emotions in the mechanism of human behaviour, the whole of literature—come to think of it, the whole of life . . .

Oh dear! If feeling, sensations, and emotions are part of the machinery of our behaviour, how do their cogwheels engage. How do they engage with the rest of the machinery? How can pains and pleasures enter into a discussion of atoms and molecules, brain cells and circuitry? Molecules and circuitry provide part of our explanations for human behaviour. But the more we can understand human behaviour in terms of molecules and brain cells and circuitry, the less comprehensible become the existence of pleasures, pains, and all that.

So here is the dilemma: On the one hand, we have Shakespeare's explanations of human behaviour couched in his language, based on a shared inner knowledge of our conscious mind—a mind that makes decisions by somehow bringing together and balancing perceptions, memories, ideas, feelings, and emotions. On the other hand, we have the explanations of brain science, based on a different kind of knowledge, also subtle and complex, and requiring insight and intelligence to understand it, but couched in altogether different terms. To provide a flavour of the kind of thing we will be coming to in the next chapter, here is another quotation—a tiny piece of the explanation of how the brain computer works. You don't have to be able to follow it for present purposes, but it is an explanation of how a nerve impulse that has been generated in one brain cell may nudge another cell to affect the chances that it too will generate an impulse—pass on the message, as it were.

> Neurotransmitters are stored in the terminal knobs in crowds of vesicles ready to fuse with the presynaptic membrane and deliver their goods in a typical eukaryotic way. The immediate stimulus for *this* is a sudden infusion of calcium ions into the axon terminal caused by the arrival of the impulse—because there are yet other voltage sensitive gates in the membranes of axon terminals and which open for calcium . . . [5]

Now *that's* science, no question. But talk of resentment, jealousy, lust . . .? Will the doorman of the Great Hall of Science let us in with such folksy notions? Perhaps we will be asked politely to leave such ideas in the cloakroom along with flick-knives, cigarettes, and other undesirables ("Yes, sir, just leave them in the tray marked 'subjective rubbish' ".)

This, then, is the first blanket objection to the idea of a conscious will in the causal scheme of things: "Sorry, we don't seem to be able to fit it in." And maybe it can't be fitted into the world as it may seem to us today—a chance-and-necessity world. But is the world really like this?

"The world"—the whole world—should include subjective things such as feelings and sensations, surely? Why should there be this conventional kind of scientific politeness not, *please*, to mention subjective things? By "the world," scientists usually mean the objective world and speak as if *that* was something well understood!

In the objective world, what seems to be the case often turns out not to be. Once upon a time it seemed that you had to keep pushing things to keep them moving. But Galileo and then Newton said no, things that are moving will keep doing so forever unless they are stopped in some way. And then Newton said there was a mysterious force which kept the moon in its orbit a quarter of a million miles away and raised the tides twice a day; but Einstein said no, patiently explaining that space is curved. That is certainly not what "seems" to me, but it is typical of the advance of science. Most dramatically, in this century the fundamental theory of matter, the quantum theory, has all but given up on what *seems*. We will come back to this in chapter fifteen. To put it briefly and bluntly, quantum theory is crazy. It applies to all the fundamental constituents of our material world. It has its own rules, and they work wonderfully in describing and predicting how matter will behave on scales from atom-sized to hand-sized to star-sized. It is just that, at the atomic scale especially, the theory does not always follow the expectations that we have built into our brains about the way things should be.

The overall message, it seems to me, is this: Mind and matter are both strange, but they belong to the same world.

If our (conscious) mind and our body existed separately how could they communicate? What would their channels of communication be made of? (Mind, matter . . .?) Such a duality of worlds is generally reckoned to have been knocked out decades ago by Gilbert Ryle's famously sarcastic description of it as "the dogma of the Ghost in the Machine."[6] But I am never quite comfortable with this piece of wit. Perhaps we are meant to think that the Machine (the brain) is the only "scientific" bit, and that the Ghost had better be got rid of. But the Ghost—the conscious part of our mind—is not going to be exorcised by shouting "Fie!" in a loud voice, or with clever philosophical arguments which conclude that pain, for example, does not really exist. Let us not be afraid of the word "subjective" (or of such "ghosts").

By "subjective" I just mean "seen from inside me." I am making no comment on what sort of stuff this *me* is made of. But whatever it is, it is an objective part of the world *for everyone else*.

Now let me give you an example of a machine with a quite harmless ghost in it—a grandfather clock. You know the general idea: The hands are driven by a train of gear-wheels, there is a pendulum and escapement mechanism in the speed controller, and then the whole thing is driven by weights.

So where is the ghost? Not in the cogwheels pushing each other by direct contact, not in the escapement—these are pure Machine. But to explain the action of the pendulum and the driving weights you need strange ideas: that forces can act across space, or alternatively that objects have gravitational fields. Spooky!

And that, I think, is what feelings, sensations, and so forth, are like too. Not really spooky, but something else about matter which we had not previously concentrated our attention on. Our aim should be to do for aspects of consciousness what Newton, Einstein & Co. did for gravitation not necessarily understanding everything about them, but naming them and putting them in context. That's much easier said than done of course, but sifting what we do understand from what we don't understand would be a start.

There is a more down-to-earth objection which is sometimes raised to the idea that our feelings can affect our behaviour. It is to say, in effect: "Sorry, there isn't enough time." This objection is based on experimental estimates of the time it takes for us to become consciously aware of something and so be in the position to exert a conscious will, as well as the time it takes for a conscious decision to be enacted.[7]

Conscious actions take longer than you might think; they go on a minimum time scale of seconds. Although a good bit faster, even automatic reflexes take time. The brake lights of the car immediately in front of you go on. There is no way that you are going to get your foot on the brake within 1/10th of a second, even if you are the current world champion racing driver.[8] And you are not going to be *consciously* aware of what is going on for another half second or so. It takes about the same time again, at the very least, to make a conscious decision about what to do. CRASH!!! Crumple. But fortunately, in learning to drive you delegated your getting-a-foot-on-the-brake skills to your automatic pilot who is much quicker off the mark. The conclusion seems clear: All of what we do on a time scale below a second or two is done unconsciously, we become conscious of these actions, if at all, only after the event.

Does this spoil the idea that we are conscious agents in any sense at all?

Not necessarily. Delegating so much of our driving skills to a faster automatic pilot, for example, is not a serious infringement of our liberty. We trained our automatic pilot and were in conscious control then. We exerted our conscious minds in the past, as it were, to be applied to future contingencies. On these future occasions we exert what we might call an unconscious will of our own making. In any case, if I have an accident as a result of the incompetence of my automatic pilot, it will be no excuse for me to protest: "I didn't do it, m'lord. Couldn't have. It all happened so quickly I had no time to think." We may not be consciously in control of every detail of what we do all the time; we may largely be spectators of our actions, but this does not necessarily deny that on a longer time scale our conscious minds are in control.

Shakespeare's Iago has an interesting analogy for how, perhaps, the will works. Here he is trying to persuade Rodrigo to get a grip on himself:

> IAGO   . . . 'tis in ourselves that we are thus or thus. Our bodies are
> gardens; to the which our wills are gardeners: so that if we
> will plant nettles, or sow lettuce; set hyssop, and weed up
> thyme; supply it with one gender of herbs, or distract it with
> many; either to have it sterile with idleness, or manured
> with industry; why, the power and corrigible authority of
> this lies in our wills.[9]

Perhaps the action of our conscious minds lies not in directly causing everything that our bodies do, but in exerting marginal control here and there. Not doing much, perhaps, indeed as incapable of intricate, consciously willed action as a gardener is incapable of willing a rose to grow leaf by leaf, petal by petal; but exerting certain critical influences with long-term consequences, especially in developing our character, makes us what we *are*.

A similar analogy is sometimes made between our conscious minds and the manager of a company. Perhaps the manager usually only comes in once a week, if it is not raining, reads over a one-page summary report, says "Very good, carry on," and then goes home. But perhaps next week, after gazing into space and letting her coffee go cold, she will lift the phone and set in motion the hostile takeover of a rival firm, creating thousands of redundancies and enriching the shareholders of Lichfield Grommets—and be voted a handsome pay rise for managerial efficiency.

High-level control should not require continuing interference. It is more about observing and thinking and adjusting, and making more radical changes occasionally when necessary. To be able to make quick *tactical* decisions may be generally applauded, but quick *strategic* decisionmaking is not usually seen as such a good idea.

Perhaps we can say that while top-level control can be quite slow, the control mechanisms should run faster and faster at lower levels: A laid-back manager is OK perhaps, but there should be whirring machinery on the shop floor.

Think about what happens when you walk: Exactly where you put your feet is left to unconscious processors in your brain and spinal cord—unless the going is difficult and you literally have to watch your step. But don't try thinking about which muscles to contract. And you will have no hope of even knowing which brain and spinal cord cells are at work. They are sending impulses to each other, and eventually to walking muscles, on a time scale of thousandths of a second. And the generation of such impulses depends on molecular processes thousands of times faster than that.

Of course it might have been a good idea for conscious processes to have been a bit more zippy. All we can say is that they are not, a fact that is no doubt a clue to the physical events underlying conscious processes. And perhaps there is a clue too in the well-known experience of events seeming to happen in slow motion during, for example, a serious accident. Presumably this is because, as with the machinery in a slow-motion camera, there is brain machinery that is responsible for our conscious awareness, which runs faster in such circumstances, packing in more items of thought per second. Thus it would give us this sense of seconds being longer than usual, of there being plenty of time to become aware of details of what is going on. The implication that there is such machinery at all is itself is a nice argument in favour of the idea that our conscious mind is not a mere spectator. Why bother with a mere spectator, of all things, in an emergency?

Chapter *2*

# The two of me

*the genius and the mortal instruments*

*T*esting one, two, three ... Here's me reporting from the inside about me. What do I find? I find an Evanescent Me, a me of the moment, a perceiving, conscious, wilful me, changing all the time, deciding, deciding ..., a me that reappears every morning, a me that I suppose my brain somehow *makes* all the time while I am awake.

Underlying this is a more permanent, unconscious, Greater Me, which I think is the more intelligent of us, a kind of civil servant coming up with sensible ideas; but also a staff of technicians with amazing computing skills, such as being able to control eating big bits of toffee without dribbling, or walking across bumpy ground without falling down in a heap. Greater Me is also an office and a library too, containing all my memories filed away somehow and so is the real substantial Me you might say ...

René Descartes saw man as having a twofold existence—a body and soul interacting with each other. This kind of theory of mind has come to be called "Cartesian Dualism" (meaning

Descartes's Dualism), but it was not particularly his idea. It is an ancient belief, built into our language, that there are two parts to us: body and soul; mortal clay and immortal spirit—there are many such pairs of terms. When Shakespeare wrote in 1599 of "the genius and the mortal instruments . . . in council" he was picking up a well-known idea. Did he get it from Descartes? Probably not. For one thing Descartes was in another country. For another, he was three years old.

Descartes's great contribution was to be much broader. The imagined dichotomy within man was to become the basis for a whole philosophical system of the Universe. More abstractly, matter and mind became the two great elements. The cut, then, was not just in ourselves but in everything. You could say that in making this distinction Descartes was helping to set the scene for the development of a new kind of philosophy: Modern science. He decreed, in effect, that matter was the proper study of science; mind, the proper study of theology. This was typical of his general strategy for solving great problems.[10] Divide them up, he said, and start with the easy bits. The innumerable disciplines of today's science, the subspecialities within them, the very way in which new science is now communicated in the form of well focused papers each concentrating on some bite-sized piece of the great question of what the world is like; all this shows how we have been following Descartes's general strategy of investigation—and to great effect.

But as a theory of mind, Cartesian Dualism is out of favor among scientists and philosophers these days, so much so that one is sorely tempted to take it up. But I think not: While there is indeed an interesting and important dichotomy in our being, it is not, I think where Descartes and tradition had put it. You will understand that when I use the term "Evanescent Self," it is not supposed to sound like an immortal soul; it is more like something which can be switched on and off, like the television.

Most scientists today will say that Cartesian Dualism is nonsense, but quite a few will insist that the mind, especially the conscious mind, is not a proper subject for scientific enquiry. Why not?—unless there is indeed some unbridgable divide between mind and matter, unless Descartes's Dualism is true. Ever since Descartes, and accelerating over the last 150 years or so, the mind has been coming back into science in spite of everything; and quite recently even the conscious mind has begun to be taken seriously as well. It's about time, too. There are so many aspects of mind which are obviously connected to things which are by common agreement in the Great Hall of Science. You will find quite a crush of Nobel Prize winners now, lugging in such hitherto contraband items of discussion as conscious awareness, feelings, free will, and so on. What next! But then even Descartes broke his own rule and speculated about where the soul and the body might come together in the brain. People are not always consistent.

We cannot but see the brain as a material object, whatever else it is. We know now that it is made up of billions of microscopic cells like the rest of the body. Nerve cells, neurons that is, are the most prominent cells in the brain and in the nervous system in general. They can have curious straggly, often enormously elongated shapes allowing them to send signals over long distances. Otherwise they are not much different from other cells.

You can think of a neuron as a tiny bag, its so-called cell body, with tubular twig-like extensions—"dendrites"—sprouting from it. The dendrites are, among other things, receivers helping the cell to pick up signals from other cells. But the typical neuron has one particularly long, smooth extension. This is the cell's output cable, its "axon" or"nerve fibre" (really, it too is a tube). It is analogous in its function to a telegraph wire allowing signals to be sent a long way off to a particular destination: to another nerve cell perhaps, or a muscle cell, or often to many other cells since these fibres are usually well branched themselves. The analogy with telegraphy is particularly apt for axons outside the brain: What we call nerves are more or less just hefty bundles of axons, bundles that may be as long as an arm or a leg. Within the brain, a better analogy might be the somewhat shorter, sometimes very short electrical connections within a computer. Here again signals are being sent to precise destinations. The main difference, as we will see in a moment, is that although the signals traveling along axons can be described as "electrochemical," the axon is not like a conducting wire. There is no electric current traveling along it, but instead a more complicated kind of domino effect that races along the axon tube walls at up to 10 times as fast as a sprinter—which is fast, but nothing like the speed of electricity.

My sketch is a highly simplified, rather formal picture of a neuron with its cell body toward the left, as well as part of a second neuron to which it is connected:

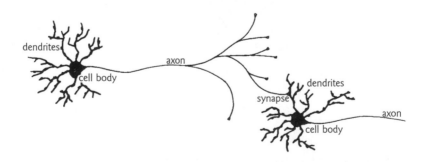

The signal would originate in the cell on the left and shoot along its axon at perhaps 50 metres a second, all the way to the little knobs at the ends of its axon. These terminal knobs can make junctions with other cells, being fixed in close contact with them, typically to their twiggy cell bodies. In the sketch I have shown just one such junction, or *synapse* as it is called, absurdly underrepresenting the complexity of the real situation: In the actual brain, a single neuron is likely to make *thousands* of synapse connections with its axon terminals to hundreds or thousands of other neurons as well as receiving a similar number of connections from the axon terminals of other cells. As a result, the densest hawthorn thicket you could imagine would be sparse in comparison to the mass of branchings and contacts between "the little grey cells" of some microscopic portion of your brain. Your whole brain has some $10^{11}$ neurons in it (that is about 20 times the human population of the Earth). In an awake brain they are all nattering away to each other incessantly.

What exactly are they saying? A good bit more than we know, I suspect, but I will stick with the "classical" view, according to which they are either saying nothing, for a moment, which can be a kind of signal when you come to think about it; or they are saying something like: "Oh go on, send a signal along *your* axon"; or else they are saying, "No, no, terrible idea. I'd keep quiet if I were you." Imagine just one of your brain's neurons with thousands of axon terminals from other cells nestled up against it whispering contrary advice. What should your hapless brain cell do in such a situation? Take a vote on it?

Well yes. Why not? It all happens automatically, of course, like this: "Yes" signals give the inside of a receiving cell a more positive *voltage*, whereas "No" signals tend to push the voltage back the other way. When the voltage reaches a critical level in the receiver cell, tiny pores at the near end of *its* axon tube begin to open up. Again, this is automatic, a direct consequence of the local increase in voltage spreading into the axon tube. Now these pores are designed particularly to let in sodium ions from the surrounding fluids, and sodium ions have positive electric charges. So you can imagine what happens. The inside of this part of the axon tube becomes more positive still. More pores open, including pores further along the tube. An explosive effect then spreads like a shock wave along the axon toward its terminal knobs. Meanwhile compensating processes behind the wavefront are already beginning to close the pores again, pumping out the sodium, generally making things neat and ready for another nerve impulse to be transmitted . . . They had better be quick about it because the next impulse may be just a few thousandths of a second away.

Having arrived at a terminal knob, how does a signal get across? How does it tamper with the voltage of another cell? The easiest way to imagine this happening is through direct electrical contact, produced by a little hole at the point of physical contact between the cells. This is sometimes the way it works. But the most common technique is through an intermediate chemical signal. On the arrival of the impulse, a terminal knob releases small molecules for a short time into the narrow synapse space between it and the receiving cell. These molecules are called *neurotransmitters*. They are tiny keys. In one type of synapse the neurotransmitter molecules simply fit key-operated pores in the outer membrane of the receiving cell and open them. That allows some positive ions to flow in from the surrounding fluids. And *that*, of course, will register a "Yes" vote because it will make the voltage of the receiving cell more positive.

You will see that if an axon terminal knob is attached to a receiving cell through this kind of synapse, it can only register a "Yes" vote if it votes at all. Synapses that vote "No" are equally set in their opinions, but work the other way around. They have different neurotransmitter keys to unlock different devices, devices designed to make the electric charge in a receiver cell more negative by letting in negatively charged chloride, for example, or letting *out* positively charged potassium . . .

This voting system is not very democratic. It's more like a shouting match, really. Neurons differ in whether they say "Yes" or "No," and they differ from moment to moment in the rate at which they transmit their characteristic signals to the cells they are connected to. And some synapses shout louder than others: They deliver more neurotransmitter. Some synapses are situated closer to the axon of the receiving cell, or are even in contact with the axon itself, and may have more effect that way too. And then timing is important: To get a neuron to fire you want a lot of "Yes" votes coming within a short space of time—very much like a shouting match. But the neuron that has been triggered into action may be one of those that always says "No." In that case, the overall effect of that bunch of yesses may be to dampen things down. Perhaps a collection of nearby cells will thus be quietened.

It is hard to see one's Evanescent Self in all this. There is little indication of conscious control here: Conscious control is much slower and less detailed than the chattering of neurons. This standard neuron activity looks more like the innermost workings of the Greater Self, which can be imagined as some kind of (unconscious) computer. The brain is often compared to a computer, but the conscious aspects of its activities—the Evanescent Self—if like a computer at all would have a very different way of working. We will be discussing these ideas later on.

In the meantime, it is not a bad first shot to say that the brain is a computer—if there is a list of *ifs* and *buts* to follow. First we should say the purpose of the brain is to be a control device like the computer in an automatic pilot, say, rather than a calculator. And although brains and computers both use simple elements in complex combinations, a neuron is not as simple as a transistor. The cells of living things are never simple: The neuron is a complex piece of machinery.[11] The signals it receives and sends may seem pretty simple, but it is clear that neurons can do more than that. For example, there are many kinds of synapses, and there is a great range of neurotransmitters operating in them. And it very much looks as though our long-term memories consist largely—if not entirely—of patterns of synapse alternations brought about by repeated patterns of use so that, for example, what has happened before may happen more easily next time.[12] This is not like computer memories, not like the current floppy disks and so on, anyway.

One must admit that for all their enviable characteristics as components of a computing system, there are some features about neurons which a neuron salesman might want to keep to himself. Although fast in relation to conscious thought, the signaling activities of neurons are immensely slow in comparison to say, what goes on in your pocket calculator. Nerve impulses travel along axons about a million times slower than electronic signals travel along wires.

Most computers today are "serial": They work by doing one thing after another in an appropriate logical sequence. Complex operations require long sequences of simple operations through which everything runs. Having to wait for one thing to be done before you can move on to the next can be a waste of time. It may not matter very much because computers can work so fast. But the brain would never get through its work like this at the speed its signals travel. The brain's trick is to do millions of different things at once so that they do not have to wait for each other.

I would like to have given you a proper wiring diagram for Greater Me (or perhaps just Greater Somebody) showing how every neuron is connected, but such detailed information is not available. There is an added difficulty: Even if all the wiring were known, its representation on paper (allowing, say, one square centimetre for each neuron) would need a foldout picture a bit bigger than the publishers were prepared to consider—some ten square kilometres. Even on that scale, the picture would be ink black with wires . . . So what follows instead are a few illustrated anatomical notes starting with the standard overall picture of a human brain, here seen from the left side:

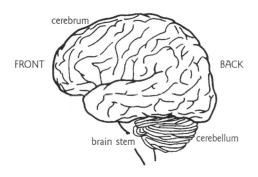

The most prominent feature is the cerebrum consisting mainly of a pair of cerebral hemispheres. Its convoluted skin, the cerebral cortex, is usually just called *the cortex*. This is quite thin, only 2–3 millimetres in most places and usually consisting of about six layers of neurons. These neurons are presumably what Hercule Poirot would refer to as "the little grey cells" and, indeed, they constitute a sizeable fraction of the "grey matter" in the brain. Incidentally, it is the high density of cell bodies here, as elsewhere, which makes it grey. By contrast, immediately under the cortex there is a mass of "white matter" dominated by nerve fibres (axons) going to and from cortex cells, connecting them with each other and with other parts of the brain.

The small brain, or *cerebellum*, has tighter convolutions. It is known to be concerned with detailed organisation of skilled actions. It is attached to the back of the *brain stem*. The *spinal cord* emerges from the brain stem without a sharp differentiation. Indeed the spinal cord is a kind of brain too. It is not just an extra big bunch of nerve fibres, but a set of computers organising, for example, much of what has to be done when we walk: sending "contract" and "relax" signals to the appropriate muscles in the correct sequence. Little grey cells are needed for that sort of thing, too, and are found in the core of the spinal cord. The brain plus the spinal cord is called the *central nervous system*.

I have given you these four megacomponents of the central nervous system in reverse order from the point of view both of their development in the embryo and of their evolution. The brain can be thought of very crudely as a collection of swellings at the front end of the spinal cord with the brain stem coming early, and the cerebellum and cerebrum being successively further swellings from that.

If we are to stand a chance of even beginning to understand how brains might produce feelings and sensations—the Big Question if ever there was one—we will need a little more anatomy in order to follow some of the ideas that are currently in the air. Let's look more closely at the cerebral hemispheres. Here, first, is a general view of the left hemisphere along with a view of both hemispheres in cross section. It shows the four main areas or *lobes*— "continents," as they might be better described—together with one of the more prominent grooves, the *central sulcus* which forms the boundary between the frontal and parietal lobes:

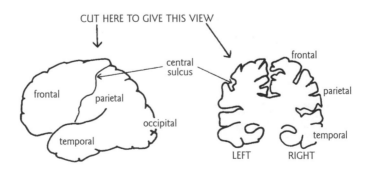

The right sketch illustrates how much more cortex there is than at first appears. More than half of it is hidden in folds, which vary in detail between individuals and between the right and left sides. The great cleft between the hemispheres is the biggest of these "hidden" areas.

The next sketch shows only the right half of a brain that has been cut in the plane of this great cleft:

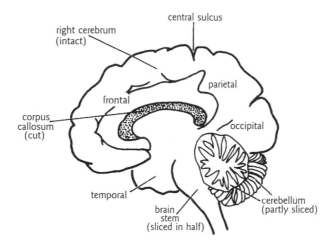

It shows the hidden zone between the hemispheres. The *corpus callosum* is a prominent landmark near the centre of the brain. This is a vast wad of fibres connecting the left and right cerebral cortex. Here it is seen in cross section.

Some parts of the cortex have clearly defined functions: the so-called "primary" regions. For example, early processing of information from the eyes is carried out in the *primary visual cortex*, the hindmost part of the occipital lobes at the back of the head. There is also a primary region in the temporal lobes for hearing. But in both of these cases, further processing is carried out in other less simply committed regions, which may be quite far away. Understanding speech, for example, depends on a region near the junction of the temporal and parietal lobes on the left side of the brain. It has inputs from the nearby *auditory cortex*, but from other places as well. This is hardly surprising since words have so many connections.

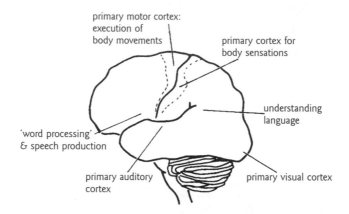

Two other primary areas of cortex are close to the central sulcus. The strip of cortex just on the parietal side of this border in each hemisphere is concerned with sensations in different parts of the body. For example, touching the left side of your tongue will cause signals to be sent to this sensory region in the right hemisphere near the bottom of the strip. Somewhat higher up are regions of the sensory strip that have to do with the lips, nose, etc; then higher up still, the hands, and so on. The bizarre convention is followed that the right hemisphere deals with the left part of the body and vice versa.

Close by, on the forward side of the border just into frontal lobe territory, there is a similar region in both hemispheres similarly laid out but concerned with actions, not sensations, with controlling the muscles of the body. This strip is the *primary motor cortex*. Again, there are adjacent and more distant, less committed areas concerned with the complex aspects of muscle control. For example, located on the left side of the brain, close to those places on the motor strip through which the tongue, lips, etc. are controlled, is the area for producing speech, which is connected to a language-understanding area further back.[13]

Still further forward of the central sulcus are the more remote and abstract seats of action. People with damage in these frontal areas can become very tiresome to themselves and everyone else. They typically lack the will to do anything, or they may be unable to sequence their actions appropriately or lack any sense of social responsibility.[14]

As a rough rule of thumb we can say that the parts of cortex behind the central sulcus deal mainly with perceiving, while the frontal brain deals mainly with *doing*. But as I say, this is only a rough rule.

Now what about this idea of an Evanescent Self? No special location has been found for it. Perhaps the idea of consciousness as a distinct part of the brain's control machinery will become redundant as we learn more and more about areas of the brain that deal with perceiving in different ways and doing different things; and as more is learned about which places become active when we have experiences of different kinds. There are the means now to see this. In the next chapter we will be talking about machines that can display the image of a living human brain on a screen, highlighting places where the local blood flow has increased for a time—presumably indicating where the turnover of energy and materials has increased, where the action is. It is like watching a nation preparing for war by noting, say, troop movements and the production of strategic materials without really knowing the detailed plans being hatched. It is like watching a brain thinking, and even what very general things it's thinking about.

We are a long way yet from a mind-reading machine, but it seems less fantastic than it once did. A broader view of the whole development of brain science during the 20th century has encouraged another optimism: that the brain is a machine whose workings we *just about understand* in the sense that we will sooner or later be able to explain it completely in terms of nerve cells and their connections, nerve impulses, bloodflow regulation and so on—briefly, in terms of current science—and that as part of this progress, the very idea of consciousness and its problematic, seemingly "nonphysical" feelings and sensations will simply evaporate. Such evaporations have been common in the history of science. One that is not so far back was a theory of life developed primarily by the French philosopher Henri Bergson and championed by, among others, Irish playwright Bernard Shaw. The idea was that living things are distinguished by having a special "nonphysical" force operating within them, the so-called Vital Force. Well, this idea of the existence of something extra and nonphysical to explain living things evaporated from discussions of the nature and evolution of life. It was never disproved, but it became obsolete with the increase in our understanding of life—what we might call The Life Revolution—brought about on the one hand by developments in the theory of evolution, and on the other by the ability to study living things at the molecular level, that is, through the development of biochemistry and molecular biology. This all happened over the last 150 years or so, the mist finally clearing quite suddenly in the middle of the 20th century as we started discovering the real molecular machinery of life and understanding it in detail.

So is that what The Mind Revolution will be like? Will terms such as feeling, sensation, and consciousness evaporate as the Vital Force evaporated: no longer used because no longer needed?

There are wise people who speak like this, using this very analogy; forgetting or ignoring, it seems to me, our shared secret that we have feelings which affect our behaviour. Vital Force was just an idea—unreal, a misunderstanding. But feelings and sensations are real. Agonising pain is not a misunderstanding. The analogy lies short.

Chapter *3*

# Qualia en croûte

*such stuff as dreams*

*I*t was a little extravagant of us to have chosen that special Michelin restaurant, but, well, when in Périgord . . . Now the moment has arrived as the dish is served. It is the dish for which the restaurant is famous, and which we have been dreaming about all day: *caille en croûte*.

A pretty enough pastry case, if a little bland looking. Pierce it gently, prise it an inch, get sight of warm soft colours. The quail meat is moist, perfectly cooked no doubt, the bird wrapped in bacon with some pâté filling the space between it and a soft inner layer of pastry. What is that scent? Truffle? Chervil? Basil? Too complex to be sure. Break it open a bit more. Even with knife and fork one begins to feel some of the complexity of texture which the French so well understand as a component of serious food. Now to taste it . . .

Across the table I see a similar concentration on such important matters. A dream come true? Anyway, a dream.

37

Some might say more like a nightmare, for what is there is not so much a quail as qualia . . .

> **Qualia,** *kwah-lia* or *kway-lia, n., plural:* feelings, sensations, emotions, moods; the qualities of our inner experiences.

And by *there,* of course, I don't mean on the plate, but inside another hard crusty case. Inside my skull. How does science deal with the stuff of dreams? Yet all of it, the pleasure of anticipation, the first sight of pastry, the soft aromas, the view out the window across the wooded hills: All of this is the stuff of dreams. It is all locked in our heads, part of our evanescent conscious selves. We *are* such stuff as dreams are made on. Our Evanescent Selves are anyway.

Just as the Greater Self can be described approximately as the computing aspects of the brain's activity—needing axons, synapses, dendrites, and so on—so the Evanescent Self is a description of the conscious aspects of the brain's activities; and *that* is all about qualia, needing goodness knows what in the way of hardware, but something for sure . . .

I hear a querulous interruption: "But what about awareness? Surely the essence of consciousness is simply knowing what is going on around you."

I have been using the words conscious and consciousness without defining them, haven't I? And I know people get upset about this kind of thing, but I'm afraid I can't give you "the definition of consciousness" because there isn't one. What a word means depends on how people use it, and "conscious," like most words, has several somewhat different uses. I can only tell you how I use the word in this book. I am supposing that to be conscious is to have feelings and sensations, and emotions and moods associated with them. Consciousness is this set of subjective effects which appears every morning. It is what general anesthetics eliminate.[15]

"Conscious" and "aware" are often taken to be synonyms. But I think this admittedly common use of words is a tactical mistake if we really want to understand what consciousness is, what is most characteristic about it. It is a tactical mistake to say that aware *means* consciously aware because it conceals the twofold conscious/unconscious nature of our being which we have been trying so hard to untangle. *Aware* can be understood, in a simple-minded way, to mean having current information on things going on within and around us. We can thus be unconsciously aware as well as consciously aware, and I will try to show you over the next page or two that this is an idea that fits our experience.[16] Now, if we agree that awareness can be unconscious, it cannot also then be the defining characteristic ("the essence") of consciousness. So let's see.

A dramatic example of unconscious awareness shows up in cases of "blindsight."[17] Someone with a damaged visual cortex may be blind in part of their visual field and deny being able to see anything, say, to the right of a point they are fixing their attention on. Asked to identify a letter of the alphabet flashed on a screen in this area, they are likely to respond, somewhat testily, that no, of course they cannot identify the letter. They are likely to be even less patient if asked to make a *guess* at what the letter might have been. But amazingly, if persuaded to try, they will likely guess correctly a good bit more often than would be expected on chance alone. Apparently there are pathways in the brain which lead to conscious awareness and others which lead to unconscious awareness.[18]

It is said that a commercial for Acme Sudds, flashed on a screen for too short a time to be perceived consciously, may nevertheless lead to increased sales of Acme Sudds. There is a law against such "subliminal advertising," so maybe it works. In any case, there are now several well-established examples of subliminal perception in normal people, where the exposure of some stimulus at a level too weak to be consciously perceived can nevertheless have effects on subsequent behaviour.[19]

Such examples should not be regarded as out on some limb of exceptional cases. They reflect something general and usual about human nature and a recurring theme of this book: Most of what we do, we do unconsciously. In walking across a room or riding a bike down the street, you have to be *aware* of all kinds of details to accomplish these feats successfully. But you are not continually muttering things like "Whoops! I'm beginning to fall over sideways. Better do something about it"—are you? There is no *conscious* awareness of such tiny crises, but they are going on all the time. There has to be detailed awareness of them—and then, of course, appropriate detailed responses too, which also are mostly if not wholly unconscious.

Not only does awareness come in conscious and unconscious forms, but these may grade into each other in such a way that the nature of the awareness does not seem to change very much, except that it becomes more conscious or less conscious. In learning a skilled act, such as eating with a knife and fork or juggling with clubs, we become progressively less conscious of details. This is not to say that we have diminished knowledge of them, that we are failing to be informed about the details needed, say, to fork a morsel of potato without breaking it or sweeping it off the plate, or that we have not processed the fact that the trajectory of the club is a little farther to the left this time and the handle is farther down . . . Quite the reverse. Surely a skilled practitioner has *more* mastery of such details and can take in more of what is happening, but does so largely unconsciously.

Nor for that matter is thought *the* defining feature of consciousness—and for similar reasons. Yes, conscious thought is a very important kind of thought just as conscious awareness is an important kind of awareness; but there are unconscious processes—intuition, for example—which are so similar to conscious thought in their effects that they are most simply described as unconscious thought.

Again there is a grading and mixing between conscious and unconscious forms. Much of what we would call *conscious* thought is in fact unconscious—when one thing reminds you of something else, for example. This is very much part of conscious thinking, but sometimes there is a delay: "Now let me see, what does that remind me of?"; and if you are anything like me, you may go on to say "Hang on a minute, it'll come back to me." The answer does come a few moments later after talking about something else. When that sort of thing happens, it seems obvious that conscious intention and unconscious retrieval mechanisms are collaborating.

Or suppose you are doing a crossword puzzle. The clue for 1-across is "Two articles measure another arrangement." You already have one of the letters:

$$. . . . . . M$$

and then later on you fill in another:

$$A . . . . . M$$

and then another:

$$A . . G . . M$$

By this time even *I* might "see" the word—anagram—and again get a strong impression of conscious and unconscious processes in tandem.

So much for *thought* as the essence of consciousness, too.

So what *is* the defining feature of consciousness? What is the difference between the conscious and unconscious forms of, say, awareness and thought? The answer is, I think, that in conscious awareness or conscious thinking the dish is being served "with qualia"; in the corresponding unconscious processes, without. Qualia are the stuff of consciousness, what our Evanescent Selves are, in some sense, *made* of.

The Edinburgh philosopher David Hume said much the same some 250 years ago:

> For my part, when I enter most intimately into what I call *my-self*, I always stumble on some particular perception or other, of heat or cold, light or shade, love or hatred, pain or pleasure.[20]

Hume saw himself as "a bundle" of such perceptions. It seems reasonable. Can *you* imagine yourself awake with no feelings, no sensations at all? Try closing your eyes to cut out all visual sensations. Did it work? You would find that the blackness still had a quality, and anyway wasn't total: There would be residual colours and lights and darks. But keep trying. Relax. Stop paying attention to any of these things. Difficult isn't it? Indeed techniques of relaxation depend on diversion, concentrating on things you are normally not conscious of, such as your breathing or the weight of your limbs. So you are not actually emptying your mind of all feelings and sensations. But it *is* possible to lose all feelings and sensations naturally, to become qualia free, even to lose the pleasant sensation of being relaxed. There is a technical term for the commonest technique of this kind: It is called falling fast asleep.

Yet I can't quite go along with Hume's "bundle." My portrait of an Evanescent Self would generally consist of exceedingly complex and changing arrangements of qualia, orchestrated no doubt by our other half: the Greater Self. Hardly a bundle.

We have been brought up to believe that serious conscious thinking is a dry, hardheaded sort of thing used for playing chess or solving mathematical puzzles. No qualia there, you might say. Yet that is not my impression, nor is it the general experience as far as I can make out.[21] Urges; hunches; unease; distaste; aha! feelings; ugh! feelings—these sorts of things enter into it. Should we call them qualia, distinctive qualities of experience? Yes, I think so. Reporting from inside my *croûte,* and from my dictionary too, I find that I can feel *satisfied* with an argument; have a *thirst* for knowledge; feel *uncomfortable* with an idea, or say that it *stinks;* or get a *whiff* of an answer. Why, one almost wine-tastes an idea when one is required to make a careful judgement on it, all of which are just figures of speech, of course. But they express acknowledged similarities between undisputed examples of qualia—perceptual sensations—and some of the elements of conscious thought.

Indeed there are cases which lie between perceptual sensations and the sensations of "pure thought." The bold brain surgeon Wilder Penfield would perform experiments on willing (and brave) patients undergoing surgery for the treatment of epilepsy. The patients were under local anesthetic and could describe the effects on their conscious minds when Penfield stimulated different parts of their brains. Penfield discovered areas in the temporal lobe, which, when stimulated, caused patients to make a sudden disconnected "interpretation." This might be an unaccountable feeling of familiarity, or of strangeness, or of things coming towards them, and so on.[22]

Appropriately timed and organised, such interpretative feelings are part and parcel of normal perception. You *see* that the car on the other side of the road is coming toward you. What do you see? The optical image of the car getting progressively bigger and bigger? Not really. The unconscious processors of your Greater Self are no doubt hard at work making calculations on that basis, but the Evanescent Self is seldom aware of such considerations. Evanescent Self just *sees* that the car is approaching as it sees the blue of the sky. It is a sensation, an "it's-coming-toward-me" sensation associated with the car as surely as blueness is a sensation associated with the sky.

Let us be more particular.

**Quale,** *kwah-lé* or *kwayle*, n., *singular*: as for example "a red quale"

This doesn't mean a measure of light waves or the pigment used to paint a London bus. Nothing as plain as that. No, red quale means the real thing, a *sensation* of red: part of what usually happens *en croûte* when someone looks at a London bus, or remembers one, or, perhaps as you are doing now, imagines one.

A hard skull, and a hard problem inside it.[23] How on Earth are such qualia produced by brains? Because they surely are. Think of those blindsight experiments, which suggest that the brain has some operating modes and mechanisms that lead to consciousness and others that do not.

Or consider the case of a colourblind painter, which has been vividly recounted by Oliver Sacks.[24] This was an abstract artist who had been particularly interested in colour—very much a colourist—and who tragically *became* colourblind at age 65 following what seemed to be a relatively minor traffic accident. From other symptoms, for example a concussion and a temporary inability to read, doctors deduced that this was due to a brain injury.

In the days following the accident, the artist realised with horror that he was no longer seeing colours: greys were odd too; everything seemed *dirty* looking. Sacks remarked that "It was as if his past, his chromatic past, had been taken away as if the brain's knowledge of colour had been totally excised leaving no trace, no inner evidence of its existence behind." Even his migraines, which had previously produced coloured visual effects, now yielded only grey images.

Someone who suffers a degeneration of the "colour receptors" in the eye (tiny cells in the retina sensitive to different wavelength bands of light) may also become colourblind. But in this case, the brain's qualia-making machinery will usually be intact. Such a person will at least be able to remember colours, imagine colours (and no doubt still have coloured migraines if they had occurred before). The colourblind painter had lost all of that. (Imagine suddenly not being able to *imagine* any colour anymore.)

There was nothing wrong with his eyes. Tests showed that he could distinguish between light of different wavelengths. This also showed that much of the brain machinery for analysing appropriate information from the eyes was also intact. What had gone missing, or gone wrong, or become inaccessible was the machinery in the brain for converting such information into a certain class of qualia—the sensations of colour.

From other evidence, we have a good idea of where the trouble is. There is a bean-sized place called V4 in each of the cerebral hemispheres near the primary visual processing area at the back of the head. More than a hundred years ago, damage to these regions had already been implicated in cases of brain-related colourblindness, but ignored. At that time, few scientists believed that there could be so precise a location for so general a part of our visual perception as colour. More recently, the colour area in the human brain has been identified even more precisely using new techniques.[25]

The PET (Positron Emission Tomography) scanner is one of a class of machines which generate three-dimensional images of the brain that reveal specific places of increased local blood flow, presumably signs of increased neuron activity. Usually there are dozens of such active areas at any given instant. But areas concerned with some more or less specific perception (or thought or task) can sometimes be pinpointed by looking for *differences*—changes in the pattern of activity when stimuli related to it are presented (or perhaps when the subject just starts to think about the topic). For example, the V4 areas "lit up" particularly when a subject with normal sight, who had being viewing an abstract painting in varied shades of grey, started looking at a highly coloured version of the same thing. V4 on the left side of the brain was more strongly active in this instance. A bisected brain sketch, this time showing the left hemisphere, gives the approximate location of V4:

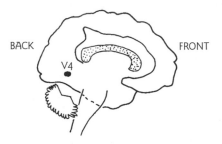

We do not know whether V4 is really where colours are made. Maybe this is just where the colour-making switchgear is. But can we doubt that, however it's done, the brain really does *make* the colour qualia? And in that case, can we seriously doubt that it makes all the other qualia as well?

There are qualia which only very few people have. Those with absolute pitch know what, say, middle C "sounds like" without reference to other notes. They know middle C as such. For the rest of us, we may not be able to say "That's the sound of middle C" as we can say "That's the colour green," but we may be good at judging relative pitch. "Sounding flat," anyway, is a familiar quale. The *planum temporale*, part of the left and right auditory cortex, usually has a fairly obvious lopsidedness, with the left one bigger. Comparing musicians and non-musicians (using another new non-invasive technique for studying live brains, Magnetic Resonance Imaging. (MRI for-short), it has been found that musicians tend to have larger-than-normal development of this region—especially musicians with absolute pitch.[26] Perhaps for people who have it, the quale for "middle-C-as-such" is made in here. But then again, perhaps this is just where the switchgear is.

The thalamus is a fairly large structure lying at the centre of the brain. Its position can be seen in a bisected view:

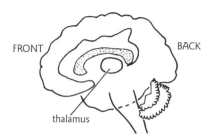

The thalamus is, among other things, a collection of processing centres or "nuclei" where signals coming from the sense organs are reorganized before being sent on to the various parts of the cortex most immediately concerned with analysing them. Signals also go from specific areas of the cortex back to the corresponding areas of the thalamus. A fair proportion of the white matter of the cerebral hemispheres underlying the cortex consists of these connecting fibres (axons). It may seem odd that the primary visual cortex should be located at the back of the brain as far away from the eyes as it could possibly be. But if the thalamus is seen as a "hub" through which signals from the sense organs to the cortex must pass, then at least this problem is solved. (Going to the thalamus and back, say, to the frontal cortex would be just as long a journey.)

If the thalamus is situated so as to be more or less equally in touch with all parts of the cortex, it might seem an ideal place for the Managing Director. And if we are thinking of the Evanescent Self in this role, then the thalamus might seem to be a good candidate as "the organ of consciousness." There are snags though. For one thing, much signal traffic that goes via the thalamus never "reaches consciousness."

Other kinds of qualia, what we might call primitive feelings—pain, fear, rage, pleasure, desire . . . —have been especially associated with central places, including the thalamus.[27] As has been typical in brain research, there have been four main ways this has come to light: noting effects of localised brain damage resulting from strokes, accidents and so on; noting effects of electrical stimulation of particular brain areas; observing changes in local electrical brain activity associated with feelings of different kinds by means of electrodes, more or less invasively; and, most recently, noting changes in local blood flow using (non-invasive) scanning machines such as PET and MRI—"functional imaging" as it is called.

The first two of these approaches recently led to the discovery of a very particular place in the thalamus involved in pain and temperature sensations.[28] In other studies, functional imaging has revealed a location associated with depression and mania. This is in the frontal lobe of the cortex tucked beneath the front of the corpus callosum, a place found to be consistently underactive in people with an inherited tendency for depression.[29]

The sketch shows the location of these and some other brain places associated with emotions:

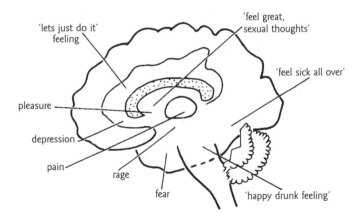

The three labels on the right side of the sketch are comments made by a patient who was wired up so he could stimulate these (among other) areas of his own brain.[30]

This cluster of central places associated with emotions is called "the limbic system." It is a wonderfully mysterious sounding name and somewhat paradoxical: *limbus* doesn't mean centre at all, it means hem or border. But then the cortex is so folded that parts of its edges come together at the centre . . .

A large and important pair of these cortical hems are immediately above the corpus callosum. These are the *cingulate cortex*. The frontal part of them, "the anterior cingulate," is an especially critical region. It evidently plays a key part in initiating conscious actions since damage in and around this area produces a state of inactivity—the patient typically loses any feeling of initiative, of wanting to do or say anything.[31] I have labeled this area as the "let's-just-do-it" feeling in the above sketch.

Another pair of hems have fancy S-shaped folds and are located lower down at inner edges of the temporal lobes: the *hippocampus*. These are also "emotional" areas although they are most famous for their role in consolidating memories. In rare cases where both hippocampal areas have been destroyed, people lose their ability to remember anything new for more than a few minutes.

Each of these pairs can be seen in a cross section of the brain sliced at an appropriate place from left to right—a viewpoint which emphasizes how these cortical edges are folded around the central region of the brain—and it shows more clearly the location of the hippocampus:

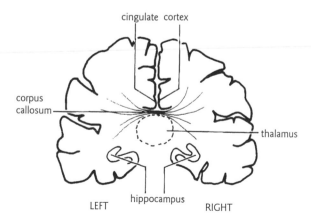

The next sketch goes back to our more usual bisected view, but with the main parts of the limbic system named and placed. Here we can see a third place where the cortex folds in toward the centre to become another part of the limbic system, labeled "lower frontal areas." This is under the front end of the corpus callosum. The stalks from the olfactory bulbs carrying information from the nose are rooted in this bit of limbic cortex.

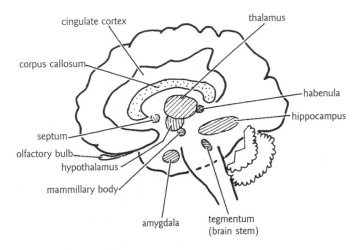

This is a very simplified picture—the thalamus and amygdala, for example, are highly complex objects in and of themselves, and I have left out the many substantial hunks of connecting nerve fibres. A wiring diagram of sorts can also be drawn for the limbic area of the brain:[32]

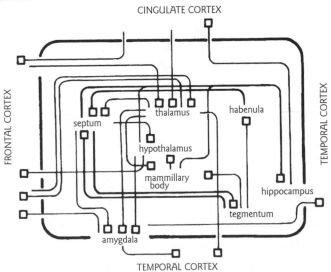

Of course nerve fibres (axons) are not actually "wires;" as we saw in the last chapter. Long-distance signals usually only pass in one direction—from the cell body of a neuron to its axon terminals, although bundles of fibres—especially big ones like the corpus callosum—will often have axons running both ways. Each of the connecting lines in the picture represents between thousands and millions of axon fibres running in the same direction. Each has a little box at one end to indicate where the cell bodies are, and hence, where the signals come from.

That is about as much brain anatomy as we will need for our discussions in the rest of the book. I will bring Part I to a close now, with a change of course inspired by these last two pictures: switching from the informative mood of most of Part I to a much more sceptical tone that will be typical of Part II. I put in these last two pictures not as a prelude to a seamless explanation of how it all works, but rather the opposite—to show how far there is to go. Do either of these little diagrams come close to telling us what fear is? Could *anything* of the sort ever do so?

Looking at such anatomical pictures and wiring diagrams, I am reminded of a toy telephone I was once given for Christmas. It worked very well. I could ring up and speak to my sister in the adjacent room. As well as the batteries running out, we ran out of things to say, and so I became more interested in checking something I didn't understand. How was it that the string between us could be slack?

I knew quite a lot about telephones because I had made one myself. What you had to do was get two empty tin cans and join them with a long piece of string by threading it through holes in their bases and knotted. Then by holding the string taut, you could speak into one of the cans and be heard through the other. I even knew in some detail how it worked: sound in one can made it vibrate, which made the string vibrate, which made the other can vibrate, which made the air in it vibrate—thus recreating sound. Armed with this knowledge I reduced my new telephone to pieces but found it completely baffling. I mean, why were the mouth piece and ear piece so totally differently from each other? The mouth pieces were full of grit and the ear pieces had neat little spools of copper wire in them. Most odd.

Was it worth destroying my Christmas present just to be baffled? I think so. My curiosity was aroused, and by the time I came across a proper explanation of how Alexander Graham Bell's invention worked, I perhaps understood it all the better for having been so puzzled before . . . But for this I had to know more about matter than just knowing about sounds and vibrations.

As I now think about the puzzle of how the brain makes feelings, a question occurs to me: Is there a similarly deep misunderstanding about how the brain really works, rooted in a still inadequate knowledge of the nature of matter?

# 2

# An Ominous Secret

## the bomb in the basement

*P*erhaps we can ignore those "open secrets" we have been talking about. Perhaps feelings and sensations are incidental side effects with no real part to play in our behaviour. Perhaps we delude ourselves: Perhaps we have no effective will of any kind, and that our conscious self is a mere spectator. And if it turns out that our consciousness does have a role, perhaps this may be adequately explained in terms that stay clear of feelings, sensations, emotions, moods, and so on. But I don't think so—and for reasons that lie within science itself. I think there is a bomb lodged in the foundations of science that may be about to go off. It is a bomb that Charles Darwin inadvertently set ticking, William James exposed for all to see—and which 20th-century thought has done its best to try and bury. **"The bomb in the basement"** I refer to is this: Feelings and sensations are part of a physical world or they could not have evolved. Yet they sit most uncomfortably in the physical world as understood by current science. It is this whole view of the world that is set to change, although whether harmlessly without injury, or with terrible loss of face remains to be seen.

# Chapter 4

# *It's a funny old world*

*there are more things in heaven and earth*

There is a kind of delusion which would surely be certifiable if it were not so common. It is the belief that the world as we perceive it is the world as it is.

Doubts about this "naïve realism" have a history going back at least to Plato in the 4th century B.C. who insisted that perceptions are only distorted shadows of reality (unless you are a philosopher, that is.) Those doubts have been reappearing again and again ever since. For example, we have Shakespeare's Hamlet being similarly sceptical about how much is known of Nature; and the Soothsayer in *Anthony and Cleopatra* speaking of "Nature's infinite book of secrecy," again as if there was a good bit more to Nature than met the eye. Some 100 years later we have Isaac Newton in a similar vein, recalling his life's work and saying how he had felt like a boy on the seashore ". . . finding a smoother pebble or a prettier shell than ordinary, whilst the great ocean of truth lay all undiscovered before me." Descartes had been radical in his scepticism, doubting the existence of anything outside himself. Immanuel Kant was to see even such basics as space and time as part of perception, not reality. This is a view yet to be taken up explicitly by science, but it is by no means out of line with much of 20th century physics. Then again, in the 20th century we have had Bertrand Russell saying that "our knowledge of the physical world is only abstract and mathematical," and Francis Crick remarking that the adage "seeing is believing" is only too true—but in the sense of your brain *merely* believing, not really knowing.[33]

Such doubts are indeed well founded. So much goes on between a quail on the plate and qualia within us. As we sit there at the table, nerve-impulse signals are being sent from the organs of our senses, mostly via the thalamus, to different places in the cortex, which don't look very different from each other. And then (we have to wave our hands a bit here) they may produce very distinctive forms of conscious awareness. Yet we know that the brain is also adept at producing similar conscious states which are not perceptions at all: remembering, imagining, dreaming or even hallucinating. That should make us wary.

Clearly as information about the world passes, say, through the eye and on to the cortex, there is a transformation from one kind of thing (light scattered off objects) to another kind of thing altogether (patterns of nerve impulses, etc.). Perhaps we think that somehow the mysterious bit—the processes that produce conscious awareness—somehow puts it right again; that it is a reversed transformation decoding back again to "reality." But this is manifestly not true. Looking at an apple does not make a real apple reappear in your brain. Of course it makes a representation of the apple. But we cannot even say that it is a "realistic" representation of the apple, whatever that might mean. That's because it consists, at least in part, of these strange things called qualia: feelings, sensations, emotions, which we *know* are not really "out there." *They* at least are made by the brain somehow. Surely all our perceptions, conscious or unconscious, are transformations—representations of reality made by an apparatus of eye, ear, nerve, brain . . ., and surely depending deeply on the nature of that apparatus.

Our private perceptions may be limited and faulty, but does science not give us a view of the world as it really is?

Well, if we were all suffering from delusions of adequacy (or, to be kinder to ourselves, illusions about the world which are adequate for everyday dealings with it), then science is in the throes of a florid delusion of grandeur. Grand it certainly is. I would not deny that it is the greatest achievement of the collective human mind. Science has certainly enlarged our perception and improved our understanding of the world, or at least aspects of it. Our sense organs have been extended and augmented—think of telescopes, microscopes, magnetic resonance machines, and so on. And our explanations of Nature have become more abstract and general. But praise can be overdone.

**THE PHYSICAL WORLD** is a useful if slippery term which means, roughly, "the world of the respectable sciences." Physics, chemistry, and biology should be included, no doubt, but the list varies according to whom you are asking. And the worlds described vary somewhat too—but they all provide representations, perceptions of reality, or to use of favourite term, *models* of reality. Almost to the last one of us, we scientists will pay lip service to the modest idea that adequate models are all we seek. Our disease is that we so easily forget. We forget the philosophy of our science and its history; and we come to suppose all over again that the present ideas of an ivy league of sciences about how Nature should best be described is the last word on the matter. That is our delusion of grandeur. Why, there is even talk of a single equation which will explain everything. (Is there a cell padded enough to contain such ravings?) No, the perceptions of science, like our private perceptions, should aspire only to adequacy. They do a good job of explaining what they set out to explain: the matter side of Descartes's mind/matter cut. But if history is any guide, this edifice called **THE PHYSICAL WORLD,** however splendid it may look to us now, will seem overrestricted and old hat by the end of the 21st century.

Again the existence of qualia provides the sharp question. Are they, or will they ever be, part of **THE PHYSICAL WORLD?**

One answer is: No.

With all this talk of not confusing perceptions with reality, how could we say anything else? Of all things, qualia are on the perception side of the fence, are they not? How can there be any connection between the ideas on which physics and chemistry are based and the feel of a pinprick? Feelings and sensations are not in same world as atoms and molecules. Anyway, you will find no serious mention of them in today's textbooks of physics and chemistry. There is no bridging vocabulary (who ever heard of an atom being bored?). Qualia exist no more in the physical world than the Queen of Spades exists in the world of chess—there is no place for them, they are banished by decree. Why, their absence is part of what we *mean* by **THE PHYSICAL WORLD,** which explicitly avoids them along with aesthetics and ethics and other "things of the mind."

A better answer is: Perhaps.

Qualia may be part of our subjective world, but they are also made by the brain, as is becoming increasingly evident. So perhaps we should say that they are *also* in The Physical World? To be caused by physical processes is surely, at least, to be halfway there, to have one foot in the physical world. Your qualia may be subjective to you, but they are, as it were, in the public domain in principle: If they are produced by physical processes, the nature of those processes might become known someday, perhaps not so far off. Perhaps the next generation of scanning machines will be able to detect and identify them.

A still better answer is: Yes and No.

No, qualia do not belong to **THE PHYSICAL WORLD** of today's science, but yes, they will belong to *THE PHYSICAL WORLD* of tomorrow. If we could look ahead to what will be understood some years or decades from now and ask if a pin (the object) and a pinprick (the feeling) are describable in the same language, then I would expect the answer to be: Yes, of course.

So I think the best answer is: Yes.

It is dark ahead. But I think that qualia will come to be treated as part of a future physical world, because the brain is a physical object, an organised collection of atoms and molecules and there is no question that it makes qualia. That is part of it. But to get both feet into the physical world, qualia must not only *have* physical causes in this sense, they must also *be* causes of other physical events.

A number of 19th century thinkers took a contrary view to this (which is still common), that the feelings and sensations of our conscious minds are (literally) inconsequential products of brain activity. For example, Thomas Huxley likened our consciousness to the steam coming from the funnel of a locomotive engine—assuredly caused by the workings of the locomotive but having no further effects on its workings.

Ironically, Darwin's theory of evolution through natural selection, which Huxley did so much to champion, tells us that qualia must have consequences for the workings of animals that have them. They must be part of the machinery of the brain, not just "emerging steam." Like the ghost of gravity which helps run the grandfather clock, the causes and effects of feelings and sensations will perhaps not be so concretely imaginable as atoms and molecules, but just as irrefutable, just as much part of The Physical World in the future. In a nutshell: Feelings and sensations cannot be inconsequential to an organism or they could not have evolved. We will follow this line in the next chapter.

# Chapter 5

# The light of evolution

*in nature's infinite book of secrecy a little I can read*

The idea that all living things on Earth evolved from simpler forms through natural selection was put forward by Charles Darwin and Alfred Wallace in 1859. It was a shocking idea at the time since it provided a possible explanation for the seemingly purposeful features of organisms and their component parts. That living things have all the appearance of having been designed had been an important part of the argument for the existence of God. Although still shocking to some people, evolution through natural selection has become the central idea of modern biology.

It was William James (brother of the just-as-famous Henry) who in 1872 or 1873 first noticed the enormous significance of Darwin's idea for "the mind–body problem" or "the problem of consciousness" as it is now more usually called.[34] If Darwin was right, said James, then on the face of it *feelings are causes*. Because if feelings evolved, then they must be useful in the struggle for existence. In that case, they must have effects in the material world. They could not be useful otherwise.

That was the beginning of it. Already it was a powerful argument against the widely held view among scientists (then as now) that feelings, although they undeniably exist, can be ignored in explaining animal and human behaviour since actual feelings have no actual effects. A quite widely held view, to which James had at first been inclined, was that behaviour was wholly due to "physiology." Feelings might come in, but it was other things going on at the same time that were the actual agents of behaviour—physiology, the working of brain cells, and the transmission of nerve signals according to well-understood principles of physics and chemistry.

James's argument against this contains the proviso: "If feelings evolved . . ." Well, did they? Feelings and sensations (or qualia as we now call them) might not have *had* to evolve. Maybe feelings and sensations are everywhere—in stones and clouds and heaps of rubbish; and that we just don't happen to know about the ecstasy or distress of a heap of rubbish because it has no way of showing it. Maybe feelings and sensations did not have to evolve—in the sense that brains did not have to evolve the means of making them—any more than, say, the means of manufacturing water would have had to evolve in organisms that were swimming in the stuff. Maybe qualia are fundamental elements of the Universe, sitting around everywhere, ready to be made use of. Who knows? It may all seem rather unsatisfactory.

But there is a second part to James's argument for the efficacy of feelings, which is independent of the question of how feelings (and qualia in general) are made, or whether they are easily come by or not. *However* they are made, and *whatever* their physical causes, we know from direct experience that they come in many different kinds. The sensation of blue is different from the sensation of a toothache, or the feeling of the carpet on bare feet, or the smell of a wine, or the taste of mustard . . . Our feelings and sensations are highly organised and, on the whole, appropriately organised too. Even if feelings and sensations themselves did not evolve, even if they are, say, incidental consequences of all material processes, the detailed appropriateness of the feelings for conscious beings like ourselves calls for an evolutionary explanation. It all seems so well designed—the way certain feelings and sensations, appropriate feelings and sensations, have been built into us to present themselves in certain circumstances: hunger, lust, feeling frightened, feeling angry, having a painful sensation. How such components of our being were chosen and organised would still need an explanation, an evolutionary explanation, even if the brain did not actually have to make them. How could it matter *what* feelings we had if they were always incidental and inconsequential, as Thomas Huxley and others had been saying: if they could have no effects on the world? Pain is a particularly potent example. It seems so obvious to us from our inner view of it that the feeling of pain, the nasty horrible feeling itself, is effective in influencing our behaviour; that the brain producing and reacting to pain is part of the *machinery* of our behaviour. James was all for taking this sort of everyday insight at its face value.

But to say that qualia are part of the brain's machinery amounts to saying that qualia are part of The Physical World. This is such a startling conclusion that it is worth being sure that we have got it right; and so I will now put the argument for qualia being part of The Physical World as clearly as I can—in the form of three not-very-startling propositions and two conclusions.

This whole scheme I will be referring to subsequectly in this book as "the evolutionary argument." Its three propositions are not outlandish. Many scientists, perhaps most, would agree with the spirit of them, but they are by no means self evident and they will need some commentary—which will occupy us for the rest of this chapter. But first, here is the formal argument:

**Proposition 1:** *Evolution through natural selection is the only "engineer" in Nature*
plus

**Proposition 2:** *Qualia are "engineered"*
gives

**First conclusion:** *Qualia evolved through natural selection*
which with

**Proposition 3:** *Evolution through natural selection only works on things which are in the physical world*
gives

**Second conclusion:** *Qualia belong to the physical world*
to which we might add:
*. . . as this world is understood now, or as it will come to be understood*

Proposition 1: *Natural selection is the only "engineer" in Nature.*
"Natural selection" describes a process in Nature analogous to the "artificial selection" used by breeders to improve their stock. Both depend on heredity, so a word or two on that first.

The main phenomena of heredity have been known for centuries. Among living things, like begets like, although usually not exactly. This allows, say, dog breeders to alter their stock in particular directions by repeatedly selecting puppies to keep and breed from through many generations. Of course they are not going to select poor specimens; nor, more brutally, will Nature. On the whole, unfit animals don't survive in the wild, and the same goes for plants, bacteria . . . All forms of life are subject to this negative selection—"natural rejection"—which helps to weed out defects, including those which are inheritable.

What those two English naturalists, Charles Darwin and Alfred Wallace, had realised independently in the middle of the 19th century, was that Nature should not only tend to weed out the manifestly unfit, the argument can be taken further. If a dog breeder can produce "improved" animals through artificial selection ("improved" according to the breeder's lights) then surely something analogous will happen naturally. Nature will not just tend to reject duds, but will tend positively to favour some variants over others; in particular it will favour variants which are *better* than the norm in the sense that they are better adapted to their environments, even if only slightly. More precisely, Nature will tend to select those which are better at surviving to produce offspring through which the beneficial features are likely to be passed on. Of course "better" means "better in the circumstance," and which variant features are to be selected will depend in general on what the environment is like. It would seem then that a variety of favourable adaptations must tend to accumulate as generation succeeds generation. The mechanisms of life will become more effective, better "engineered" for their purposes.

Sexual reproduction complicates this story while making it more plausible. The motto here is "match & mix." Through sexual reproduction, characteristics which have evolved independently in different lines can come together. Thus evolution need not be a first-do-this-then-do-that sort of business which would make it very slow; it is a parallel process too. Different research and development lines can, as it were, be pursued at the same time in different places.

We can see then that there are three key features of heredity that allow natural selection to operate:

Like tends to beget like.
There are sometimes variants.
These variants then also tend to breed true.

How does it happen? What is the physical basis of heredity? Even to Darwin it seemed a tall order to explain all this; even taller if you consider match & mix sexual reproduction where characteristics can be shuffled around to some extent and evolve in a quasi-independent way.

The main part of the answer to this question is that inheritable characteristics are determined by genes. What you inherited from your parents was not the set of your characteristics, but a set of genes which determined them. Genes are like books in that they hold information—messages, instructions, recipes; or they are software if you prefer, programs. They are written as strings of four symbols—call them 0,1,2,3—rather in the way that computer information is written as strings of two symbols, which can be called 0,1. Like information written in books or on computer disks, these messages are easily reproduced and so can easily be passed on to offspring. But the genetic message holders are neither paper and ink, nor disks coated with magnetic material. They are incredibly smaller structures: actual individual molecules; DNA molecules, although (for molecules) they are also incredibly long. A living thing's genes are its construction and repair manuals. There is a lot of "writing" in there as you can imagine (about a million pages worth for the human set of genes).

Now how can one say that natural selection is the *only* "engineer" of evolution? I do not mean to suggest that natural selection is the complete explanation for evolution, but it appears to be the explanation for that most extraordinary feature of the process of evolution: It seems to mimic the activities of an engineer and give rise to contrived objects, *machines* of one kind or another.[35] Start thinking about any part of an organism—a leaf, a wing, a hand, a liver . . . —and you will soon be thinking about what it is *for*. This is not at all the sort of question we would ask about the moon, or a mountain, or a crystal of sugar. Such things lack that vague but essential feature of living things that they *seem* to have built-in purpose, to be "engineered."

Much of the secret of this crucial evolutionary feature can be found in a kind of long-term memory that living things all have. They can, as it were, build on success: they remember, as it were, what had been successful before. In organisms that are reproducing fast enough, blind trial and error provides a successful means of improvement because, under these circumstances, the errors don't matter. The successes live on in endless copies of copies of copies . . ., and thanks to that amazing long-term DNA memory, the good idea once hit on can "live" for millions or billions of years and be built on,—and built on, and built on . . . —for millions or billions of years.

The famous bacterium *E. coli*, supposedly a simple form of life, has been evolving for billions of years. It is far too small to see with the naked eye, and yet it is far more contrived than a jet engine— no joking.[36] As for brains . . . Well, they too are products of evolution through natural selection. Nature is way ahead of us here.

Proposition 2: *Qualia are engineered.* Feelings such as hunger and lust have obvious parts to play in surviving and producing offspring. They are well adapted for this, contrived for these purposes as are countless other such elements of consciousness. But, as we saw, two questions have to be disentangled. How is it that qualia arise? and: How is it that they are appropriately connected up?

The simple brute fact of the existence of, say, *pain* in the world cannot really be explained by evolution, not at the deepest level of how such a thing can exist. That has more to do with physics, with the nature of the stuff that the Universe is made of—"stuff" we will be talking about in the final chapter. But recall that there is a more superficial level that needs explaining too: an engineering level, the level of appropriate arrangement. Even if we must simply accept that there are such things as pains and pleasures, and that we do not know how they arise from the material brain, and that we thus have no way of saying whether or not any kind of "engineer" would have been needed to *make* them—even then the question remains: How did the qualia of our own conscious minds come to be so appropriately chosen, so appropriately wired up? Here, at least, some kind of "engineer" would have been needed, leading to our interim conclusion that, yes, qualia evolved as the machinery for making qualia evolved as the genes for making the machinery for making qualia evolved. The language is the same as for anything else that can evolve through natural selection.

Proposition 3: *Natural selection works only on things which are in the physical world.* "Things" is a deliberately casual word. It could be used to cover, say, the flight muscles of a fly, or the neurons which control them, or perhaps aspects of the fly's eyesight or the wiring of its visual circuitry—all of these being things that are good for a fly to have to avoid being swatted too easily or eaten by frogs. "Being quick off the mark" would be the Greater Thing, but it too would evolve as its sub-Things evolved and became more suitably coordinated.

The well-honed excellence of a fly's flight muscles might have evolved in the following sort of way: Because of the frogs, and fierce people with good swatting techniques, flies that had slightly better flight muscles would be more likely to survive and have offspring in the form of dear little maggots. When they grew up and became proper flies themselves, they would be that much more likely to have the improved flight-muscle design. The reason for that might be that they had inherited some moderately long piece of DNA instructions, which was responsible for controling the manufacture of the flight-muscle protein and which had some (probably tiny) alteration in these instructions. "Mutations" like this would have appeared originally by chance, perhaps from some slightly careless copying at some point in the past.

Most mutations are harmful. The most damaging ones will be eliminated at once, and many are harmful enough to be eliminated quite soon. But no matter. Flies breed like flies, and sooner or later beneficial mutations will appear by chance, and some of these will catch on.

For a fly, this skill—this Thing of "being quick off the mark"— has physical causes (message molecules inherited from Dad and Mum), and it also has physical effects (improved chances of survival). But you will see that the argument is general. Things can only evolve through natural selection when they are doubly causally connected like this, when they have physical causes and physical effects. Well then, to have physical causes and physical effects, what more could one want as entrance qualifications to the physical world?

So we conclude that since qualia evolved, they belong to the physical world.

And this, I think, is a bomb in the basement of science, because qualia do not at all seem to be describable in the same language we use for atoms, molecules, physical motions, or physical force . . . Descartes might be blamed for having carelessly left this bomb in the first place when he divided the great problem of the Universe into the problem of mind and the problem of matter.[37] Made with the best of pragmatic intentions, to divide a problem into more manageable parts, it quickly came to be seen as an objective fact about the Universe rather than a sensible strategy of enquiry. Well, Darwin and Wallace set the bomb ticking and the young William James exposed it for all to see with his evolutionary argument for the efficacy of feelings. Unfortunately, the 20th century behavioural- and neurosciences seemed more impressed with the title of an essay that James wrote much later—"Does 'Consciousness' exist?"—and its assertion that consciousness is not a thing but a process.[38] Ironically, this was written in 1904, one year before a young man named Albert Einstein was to remove any depth from the distinction between "thing" and "process."[39]

So here is the final conclusion (now extended a little): *Qualia belong to the physical world, and if that does not seem to be the case then we do not yet know enough about the physical world.*

Will soft words yet defuse the bomb?

Chapter *6*

# *Straight talk, double talk, fast talk*

*this is a very false gallop of verses*

We learn early in life that two excuses are always worse than one. Even if they are only *somewhat* fictional, two stories can tangle, and inconsistencies are easily detected by the alert human brain. A second reason why one should never give two reasons for anything is that the alert human brain likes things to be simple. Perhaps this is because of its dismally limited conscious thinking capacity, but there is no doubt that we do like a straight answer even when perhaps there isn't one. We want to know *the* cause of that thing that happened yesterday and upset Mrs. McGinty, and the *real* reason why she was so annoyed.

Yet of course there are many circumstances where reasons and causes come in droves. Accidents *always* have multiple causes, events which if only they hadn't happened just the way they did would have led to no accident instead. If only the truck driver had drunk a second cup of tea with his breakfast, as he usually did, he would not have been coming around that corner just then. (If only last Thursday the shop hadn't run out of his favourite brand of tea

71

. . .) Historical explanations like this quickly become hopelessly complicated and boring. It's much better just to blame someone: "It was Jimmy's fault—and don't start making excuses like he has a wonky gene for spatial orientation or that it was his wretched up-bringing that turned him into an alcoholic so that he had his usual hangover that morning and was slow to see how big the truck was and didn't leave enough room for it . . . He screwed up, right?"

We also like explanations to be simple, but they often aren't. "Thou shalt make things as simple as possible" is a scientist's first commandment, to which Einstein wisely added ". . . but not simpler." One should be warned, however, that not only is everyday life infernally complicated, but so is biology. Biology is far more about complexity than simplicity. I mean, just think of evolution, or think of the brain.

But the question I want us to think about now is: When is it legitimate and when it is not legitimate to have more than one explanation for the same thing?

Remember Iago and his neurotransmitters? Was it feelings of resentment that made him so beastly, or was it all those telegraph signals flashing in his brain? It may seem to be double talk to say "both." But it is a straight enough answer if we mean "a bit of each"—that they were *contributing causes*.

Contributing causes are fine, but there is another kind of "both" answer that is popular among superstitious readers of astrology columns in the newspapers. Our major life events are explained in terms of the positions of the planets at our birth, while at the same time it also seems to us that more everyday explanations of why things happen must be true too. (Surely it is clear: A truck driver not having a cup of tea might lead to our early death.) If the rest of us think that astro-illogical explanations are crazy, it is perhaps because we sniff heresy, what is called *double causation*:

"You can't have it both ways" is the motto in this case. Two totally different and complete-in-principle explanations cannot simultaneously be true. They cannot even sit together, never mind gallop.

All this sounds perilously close to what is perhaps today's scientifically favourite way of resolving the dilemma of Iago and his neurotransmitters. It is yet another way of replying "both," but saying something like this: "Yes, Shakespeare's account of Iago's behaviour in terms of his and other people's feelings is OK, but so would an account in terms of neuro-electrochemical computing if only we knew all the details. There is no clash because they are *equivalent descriptions:* They are just different ways of saying the same thing."[40]

There are plenty of nonproblematic examples of "different ways of saying the same thing." Why did the cork pop out of the ginger beer bottle on the sideboard? I might try an explanation like this: "Well, you see, these collections of atoms called molecules are very tiny, far too small to see but you can take my word for it that they are in perpetual motion, wobbling, spinning, flying about, and colliding with each other and with things around them. Sometimes they stick together, firmly in solids, more casually in liquids. But in a gas such as air they move freely. Molecules of air in this room, for example, are colliding with every surface in it. Now where was I? Oh yes: Before the ginger beer started to ferment seriously and produce lots of carbon dioxide gas, there had been about as many haphazard collisions per second between air molecules and the top surface of the cork as there were collisions between gas molecules inside the bottle and the cork's bottom surface. Each collision would give a tiny push to the cork. Collisions from the outside would try to push the cork in, those from the inside would try to push the cork out. But the net effects in each direction were sufficiently well balanced for the cork to stay in place. Once fermentation got underway, the extra gas being produced inside upset the balance. Now there were a lot more molecules pushing from the inside, and the cork flew out and frightened the cat.

Or you might say that the cork popped out because the pressure inside rose too far above atmospheric pressure. The first explanation was more detailed. The second is more convenient. But they are equivalent explanations because we know what is going on. Pressure just *is* a net effect of molecular collisions and it is easy to see how.

There is no such transparent equivalence between feelings, etc., and a net effect of neuron activity in the brain, however.[41] To say that we have an equivalent explanation here is simply not illuminating. We can only say "maybe so," or as I would prefer to put it, "maybe not."

In fact, come to think of it, probably not. It all goes back to "the light of evolution" and satisfactorily answering the question of why our qualia should be well adapted. I mean, it just *is* the case, is it not, that feelings such as hunger and lust, our various sensations of pleasure and pain, of enjoyment or distaste, can be ascribed to what is good and bad for us and our kind? They are on the whole more likely to lead to our survival and the propagation of our kind. Now if such qualia were simply alternative descriptions of a situation that has a full underlying explanation in terms of neural computing—in the kind of way that the idea of *pressure* give us an alternative way of describing what molecules can do to ginger beer-bottle corks—then the appropriateness of our feelings would be inexplicable. It would be inexplicable in evolutionary terms because feelings would not actually be *doing* anything to explain why they had evolved, and why they are so sensibly wired up. Natural selection is not interested in things that are simply alternative descriptions.

Several well-known features of brain activity support the idea that conscious experience is connected to activities of neurons, but not the idea that conscious experience is *just the same as* the totality of neuronal activity at any particular moment in somebody's brain. Feelings and other qualia are produced by the brain; but they are produced only sometimes, and in an immense variety of forms, more or less intensely, and on different occasions. This gives all the appearance of qualia being the products of distinct sets of systems in the brain which may or may not be switched on, and may be set at different levels—while corresponding unconscious perceptual processes, unconscious thought processes, and unconscious processes involved in precipitating actions are proceeding at the same time. We have discussed this sort of thing at some length—how we can be more conscious, less conscious, or unconscious of what we are doing as the same general activity passes from novelty to familiarity. PET scans may show a changing pattern of brain activity as this happens: activity strongly focused in particular places to begin with, and then becoming weaker with the foci changing as a novel task becomes better understood and no longer requires careful conscious attention.[42]

Carbon monoxide poisoning happens quite often (frequently as the result of a faulty gas fire, would you believe it?). It is something to be avoided at all costs. It can produce death, or worse— massive dehumanising brain damage. But sometimes it causes selective damage to particular mental functions. In one case, a young woman was found to have localised brain damage, especially in regions close to—but not in—the primary visual cortex. At the same time she had developed a severe perceptual defect: being unable to recognise objects by sight, even though her visual field was intact. She was, for example, unable to distinguish a vertical line from a horizontal one; she could not tell whether a postal slot was vertical or horizontal, for example. Yet when asked to post a card through such a slot (orientated differently on different occasions), she would do so skilfully. The conclusion derived from this study was that the brain-processing systems underlying conscious perception are not exactly the same as those required for automatic skilled actions.[43]

There have been similar reports of the "duality of motor function" in cases of stroke victims with localised brain damage: they may be unable to carry out a simple, explicit verbal instruction to perform some action, say, to lift a teacup, but then a few minutes later when offered a cup of tea, they will do so easily—"without thinking."[44]

One might be tempted to say that when brain damage has occurred and it is possible to find out where it is localised, and when some loss of conscious function has taken place at the same time, then the location of that conscious activity has been established. I am inclined to think that this could well be so, but there is the danger of fast talking here. Recall the question we discussed in chapter 3, of whether V4 is where the colour qualia are made. That posed a similar difficulty. Maybe such places are where conscious phenomena are controled but not actually where they are made.

Not to go too fast, we have to be careful to distinguish between three ideas: correlation, causation, and identity.

If I push a button on the wall outside our front door, a bell rings in the hall. Here, there is causation between the two events but not an identity. I may be inclined to say that "I rang the doorbell" but what I really mean is that I pushed a button which closed a circuit which allowed an electric current to flow in a wire leading to the hall where it energised an electromagnet which pulled a wiggly piece of metal toward it which had a hammer attached to it which hit the bell—'Bing!'—but at the same time broke the circuit so that the wiggly bit of metal sprang back again and the whole process was repeated 'Bing-Bing-Bing-Bing . . .!' " So I shouldn't say "I rang the doorbell." I should say that I pushed the door button which caused the hall bell to ring. Even that is going out on a limb a bit. Being really cautious, one should say only that button pushing *correlates* with "Bing-Bing-Bing-Bing . . .!"

OK, so that is pussyfooting: We know there is a well-understood causal chain of events here, anyway somebody knows what it is, so we might as well talk causes rather than correlations. We use the word "correlation" when events happen together or are associated in some way, yet we are not sure whether A caused B, or B caused A or both were caused by C or . . . Well, when we really don't know what's going on. With things that nobody really understands, we have to be prepared to pussyfoot. But let's not rush into it.

I want you to think about two kinds of "initiating causes," or "stimuli," as one might call them: "pushing the button by the door" and "light from a cloudless sky entering the eye" Suppose that these have consequences, respectively: "a bell ringing in the hall" and "experiencing the sensation of blue" In each case we may suppose that there is some sort of chain of causation, and we think we know at least which way it goes—blue light causes the sensation blue and not the other way around, for example. In each case, there are mechanisms which make contrived connections. There is no law of Nature which says that pushing buttons causes bells to ring; this only happens if things are suitably contrived and wired up. It is the same, presumably, with light from a cloudless sky and the sensation blue; except that it was not an electrical engineer, but Nature's engineer—evolution through natural selection—that did the contriving. So far, the analogy holds up quite well—no pussyfooting yet.

Now let's wade in a little deeper with two innocent questions: What is identical to "a bell ringing"? and: What is identical to "experiencing blue"? I am not asking what causes these things, but what they *are* in physical terms. Recall the major conclusion of the last chapter: It was that qualia, things like experiencing blue, are "physical." They must have both feet in a "physical world" (of some sort) or they could never have evolved. So of course we should be attempting to answer the "What-on-Earth-is-it?" questions for blue as well as for bells.

Let's try bells first. The ringing is not *all* the events taking place between pushing the button and the generation of intermittent compression waves in the air in the hall. No, it is to be identified only with these compression waves, no more. The physical sound *correlates* with every part of this chain of events but is only *identical* to one part of it: the compression waves.

Now there must be a similar situation between light from a cloudless sky entering the eye and the sensation of blue. In the first place, of course, qualia are no more identical to their external physical causes than pushing a button is identical to ringing a bell, even if we may talk as if this were so, and naïve people may think it is so in both cases. Nor is an identity with consciousness to be found in the totality of the causal chain between, say, light entering the eye and the sensation of it. Not even all the neuronal activity within this chain of events could be identifiable with sensation. It is well established that not all neuron activity leads to qualia production. For example, in the visual system (as we shall see in chapter ten) most of what goes on in the brain never reaches consciousness. Then again, we too easily suppose that the qualia of consciousness are produced by something *like* the elements of the causal chains leading to it (nerve impulses and so on). This might be so, but to imagine that it *must* be so is like saying that because the causal chain from the bell-push to the electromagnet in the bell mechanism is electrical, it follows that sound must be a kind of electricity.

Why do so many people seem to imagine that qualia have to be identical to (some) "normal" neuron activity? Caused by such activity? Yes, in part at least. Correlating with it? Certainly. But *identical . . .?* One might suspect here a misguided attempt to make things simpler than possible; a much too hasty and mistaken identification of something we do not understand with something we pretty well do. As we have been discussing in this chapter, such an identification does not fit with the idea that qualia evolved, or that brain activity changes with consciousness; nor for that matter does it fit with common sense. Sending signals and making qualia are on the face of it, well, *different* and might be expected to call for at least somewhat different underlying hardware and/or at least somewhat different kinds of activity.

Straight talk, double talk, fast talk . . . I have been trying to persuade you that the real reason for Iago's behaviour is to be found neither wholly in his brain's neuron signaling, nor wholly in his strong feelings. Either of these attitudes is just lazy fast talk trying to make things simpler than possible; nor is it to be found in a premature identification of feelings with nerve impulses. This is a case of mistaken identity. There is no such simple superposition, it seems to me. And it is a kind of double talk—and lazy too: Neuron signaling and sensations are not just "different aspects of the same thing," as many would have us believe. I have been trying to persuade you that the straight talk is to say that human behaviour comes from the combined effects of two distinctive modes of control, that "a bit of each" is the appropriate motto.

# 3

# Secrets Unfolding

## all in a brain's work

*A*fter the three sceptical essays of Part II we return to the actual brain to see if we can begin to see how anything as apparently unphysical as a moment of fear or joy, or a feeling of uncertainty or agony could be part of the explanation of how a very physical brain operates; when most of what it does is cold calculation, and where there is no obvious reason why it should ever do anything else. But now we have the advantage of abandoned hope: There will be no complete solution to the problem of consciousness couched in the language of current science. Or, if you prefer, we have the added excitement of an impending expansion of science. That can now be in the back of our minds as we restart (after the breakdown at the end of chapter three) and return to specifics, beginning with a preliminary conjecture on the nature of the machinery that makes feelings (for there should be such machinery). Then we will talk of drugs and other switches of our conscious states; and make thumbnail pictures of how the conscious and unconscious collaborate; and catch glimpses of the incredible unconscious analyses and calculations that underlie perception, with visual perception as the prime example. Indeed, there is a monster out there in the dark, but its shape is emerging.

Chapter *7*

# How might *brains* have feelings?

*those musicians that shall play to you hang in the air*

$A$ brain going about its everyday business, solving problems, getting things done, being conscious of some of it, having feelings about some of it, sometimes. How do we represent that? Here is a brain (and spinal cord) solving a problem unconsciously. It is walking its body across a room "without thinking," that is to say not consciously thinking about the walking:

problem: to get across the room  problem solved: got there

The big curvy arrow is not supposed to represent the movement across the room as such, but the detailed streams of computation needed to perform the action successfully: what computer people like to call "processing"—in this case massive, parallel, not fully understood, but not deeply mysterious either; not *so* unlike the processing that goes on in an ordinary computer.

Next, I would like you to imagine the brain of a body whose right leg has just been put into an enormous plaster cast. Here it is steering its body across a hospital ward, painfully, self-consciously, and giving great attention to every dicey move:

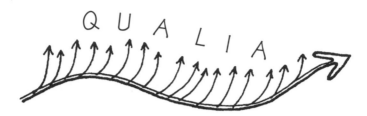

It is not just that there are, well, a few feelings and emotions in the background of the *real* things going on—a kind of background music. No, the "music" by any judgment is part of what is going on. Remember that it is caused physically and has physical effects, too, according to the evolutionary argument of chapter five—and no doubt according to the patient and the nurses looking after him too.

All of this still falls far short of an adequate description of what a conscious brain does. But even at this banal level of simple cartoons and rough analogies, there is room for disagreement. Even this picture represents only a tentative hypothesis. It reflects the conclusion of the last chapter in saying that our feelings and sensations are close to, caused by, but not simply identical to all that "neuronal computing." They are something added, like music, but not anodyne background music. In such cases there is an orchestra of qualia playing for some purpose.

But if we are to say that our conscious mind is an organisation of qualia and is a bit like music—in the way it lives in Time, in its richness and complexity, in its continually changing and inter-weaving strands—then we have to admit to some ignorance on the matter of the instruments. They seem to hang in the air. Where are they, the qualia makers? What are they? How do they work? Why, we don't even know how big they are, or how numerous; or whether the variety of qualia comes from a variety of instruments, or from one instrument that can play many tunes . . .

It is time for another definition—this time cooked up for the occasion:

> **Qualagen,** *kwah-la-jen* or *kway-la-jen, n.* putative evolved brain machinery directly concerned with making feelings, sensations, and other qualia: **qualagenic,** *adj.* describing a structure or process that makes qualia. Also **qualagenesis,** *n.* the process of making qualia.

I should now issue a warning that a nasty collision with our intuitive "common sense" must lie somewhere close ahead. Two chapters back we saw it coming. Our course was set once we decided that qualia must have evolved since it then followed, as night follows day, that they must have both physical causes and physical effects.

What we are contemplating in thinking like this and naming our quarry is just a rather sharp version of what Francis Crick calls "The Astonishing Hypothesis"—that our inner life is a result of "the behaviour of a vast assemblage of nerve cells and their associated molecules."[45]

Many people reject such an idea out of hand, perhaps muttering "If you believe that you'll believe anything." But the issue is not one of personal belief, not whether it is obviously absurd, but whether it is *true* in the good-old scientific sense of fitting the facts. Let me remind you again of Isaac Newton's *really* astonishing hypotheses, making Crick's seem rather timid by comparison. Newton said that *every* object in the Universe which has any mass at all, yes every single one, attracts every other such object across the vast reaches of space with a calculable force. He was never very comfortable with this idea,[46] and his excuse for suggesting it was that it seemed to be true. It fitted the facts. Amazingly, most people now think it is obvious. It's not: It's just that we all learned it at school and have since forgotten to be astonished by it. As a final twist, it has turned out to be not only not obvious, but actually not quite true; that is to say it does not fit the facts perfectly. Einstein produced a better idea which abandoned the notion of forces operating at a distance. To come up with something less astonishing perhaps? Not at all. The mass of an object produces a curvature in the space around it, he said—to which one can be forgiven for saying: "Ah! I see, er, *right*" and quickly moving to another topic. There's no escaping astonishment in this area.

And there's no escaping Crick's astonishing hypothesis that we are a vast assemblage of neurons, etc. But, in line with the last chapter, this is not to say that transmitting and receiving nerve impulses also makes consciousness directly, just like that, without the need for any additional explanations or hardware. So we are led to look a bit more closely at what else neurons can do, or with what else they are closely associated.

Of course, not knowing what qualia-making machinery would be like, I can't be dogmatic about where at the neuronal level the critical action might be. But if it is anywhere, then perhaps it should not be right in the middle of systems that deal with making nerve impulses. We know from our subjective experience that conscious events come and go in seconds, not in the milliseconds of neuron firing. Imagine you are a harassed "pyramidal" neuron (one of the big ones in the cortex with long-distance connections) and you are trying to help put together a nice quale. All the time Wham! Wham! Wham! . . . *huge* changes in local voltage are hitting you every few milliseconds (for some numbers see note[47]). You might want to look for a quieter place—Near the main action no doubt, connected up to it, but perhaps not right in the middle of it.

The outer reaches of dendrite trees might be quieter. Or it might be worth considering some of the smaller, less prominent neurons of the cortex. Or perhaps a more obvious idea would be to look at all the other cells in the brain.

Like most people, I have been talking (so far) as if neurons were the only cells around. But 90 percent of brain cells aren't neurons at all: they are *glial cells*. It is a rather disparaging name, meaning literally "glue cells." They are everywhere between and among the neurons, occupying the same general space.

Glial cells, which come in several different kinds, are now seen as providing not only physical support for neurons, but also a kind of valet service that generally looks after the chemical environment of the neurons. It performs such tasks as sweeping up neurotransmitter molecules from around the synapses so that "yesterday's news" (i.e., a signal that arrived a few milliseconds ago) does not continue to have effects for too long into the future. Another class of glial cells provides fatty insulation for long axons, speeding up signal transmission, and incidentally making the white matter so white. More recently, it has been found that a glial network of star-shaped "astrocytes"can do some signaling on its own. Glial cells are not just "glue." Perhaps they are not just "servants" either. Insofar as they are necessary for the efficient working of axons and synapses, for example, these glial cells, which lie so close to every neuron in the brain, would be in excellent position not only to be influenced by, but also to influence general neuronal activity.[48]

Of larger-scale structures and processes that might be qualagenic, we saw that central "limbic" regions are associated primarily with emotions and moods. And the visual region V4 seemed to do very particularly with colour sensations.[49] Brain scan experiments on conscious subjects have shown how different areas of the cortex may become active—indicated by changes in local blood supply—

when people think in different ways or even simply *imagine* different kinds of things.[50] Evidently, conscious perceiving, or thinking,
or imagining all use energy. It is hard to say how much, if any, of
this energy actually goes into qualia production itself as opposed to
background unconscious neural activity needed to support it. Since
qualia are physical effects, then of course we should be asking about
the energy required to make them, and suspecting, perhaps, that
some of these visibly increased bloodflows reflect this.

But is this the right scale at which to be looking for the
qualia-making machinery, the qualagens? How big is a qualagen? It
is well known that the brain, like other organs of the body, has a
hierarchical structure. At the bottom of the hierarchy there are
atoms. These are mostly joined together to make molecules at the
next level up and, in turn, assembled together into "organelles,"
which are more complex machines within cells. They can be
thought of as "sewing machines" for making protein molecules,
chemical "batteries" for providing power, and many other such
things all enclosed in an outer membrane to make a cell. Then
many cells join and communicate with each other to make different tissues. Cell types and structures in the brain differ from one
place to another, which is crucial to the brain's detailed workings.
And then there are the "organs" of the brain: the cerebral hemispheres, the thalamus, and so on. Different aspects of brain activity
can be understood as "emerging" at different levels of this hierarchy
as we have already begun to see.

Where in this hierarchy does consciousness come in? An obvious answer is that it must be way at the top. But maybe that's not
true. Maybe qualia are not elevated members of this hierarchy.
Maybe there are molecules within cells which, when they are
kicked into activity, tend to coordinate with similar molecules in
other cells over distances of centimetres. Maybe they play in an or-

chestra that transcends cell boundaries. Maybe the qualagens, the brain's machines for making qualia, are thus both microscopic (in the sense of being made up of molecular mechanisms), and macroscopic in the sense of occupying large areas of the brain—and that is why they have been so difficult to find.

To the extent that they are perhaps large-scale effects arising from large numbers of very small-scale processes, conscious states would be analogous to the gas pressure in that ginger beer bottle from the last chapter. This was a consequence of vast numbers of molecular processes—collisions of gas molecules. Our analogy, however, does not continue any further, since in this case the microscopic processes we are trying to imagine as those directly underlying qualia production are unknown, and they are not necessarily the same as any of the processes underlying the sending and receiving of nerve impulses.[51]

Protein molecules are likely to have something to do with the qualia-making machinery we are trying to imagine. They are made more or less directly from information in the genes; and we find that in cells of all kinds, proteins do most of the clever stuff. So let's suppose that protein molecules are at least critical components of the qualagens.

As molecules go, proteins are big. A protein molecule in a cell is likely to contain several thousand atoms connected in a particular way, which allows that protein to be a machine with some particular function. But as *machines* go, proteins are tiny almost beyond belief. To examine the innards of a Swiss watch, which used to be the standard for fine-scale engineering, we need a magnifying glass. Today, we have still-smaller man-made mechanisms, such as computer chips, which require a microscope to examine them. Now try to imagine machinery on a still-smaller scale, say a hundred times as small, needing a microscope just to see it; a hundred times smaller again and you would need an electron microscope; and then *another* hundred times smaller still and you would need the highest-powered microscopes available. Now, at last, you would begin to arrive at the territory of protein engineering, the engineering on which all life on Earth depends—of what most of life can be said to *consist*. In any cell in your body there will be many thousands of different proteins doing thousands of different things, and usually many thousands of copies of each.

Imagine now that you could open up just one of the million billion cells that make up your body and take out some of these exquisite pieces of protein machinery to look at in detail.[52] Perhaps the first protein molecule you come across performs one of the myriad of chemical transformations that a cell must bring about to keep going. Next you might find a neat little cluster of another type of protein as part of a membrane, which helps form a pore that will let through just one particular kind of ion, say, potassium. Or perhaps it's a more complex cluster of several proteins that together act as a selective pump, pushing just certain molecules across the membrane in one direction only and using a molecular fuel to do it. Other proteins give the cell a kind of skeleton which helps keep it in shape—and very odd shapes they are in the case of neurons with their sometimes dense trees of dendrites, their sometimes long, wandering, branching axons . . . It is hard work making and keeping a shape like that. One class of these skeletal proteins form "microtubules" helping, among other things, to buttress axons. These also act as railway lines transporting materials between the cell bodies of neurons and their axon terminals—and yet other proteins act as little engines that do the pushing.

In the old days people talked of "protoplasm" as the magic stuff in cells, the stuff of life. Then when the complexities of the ma-

chinery in protoplasm were discovered, there was substituted another kind of incredulity: How could machinery that is so tiny be so
complex and work so well? Well it is, and it does. We might ask
how the protoplasm of "brain matter" can be so contrived that it
makes feelings and sensations, and so reliably. Well it is, and it
does.

Even if you want to say that qualia emerge at higher levels of
organization, at the level of cells or circuits or more complex as-
semblages of such things, all are made of molecules. It would still be
true that qualia come from molecules, suitably organised.

But I am asking you to suppose that elementary feelings and
sensations might arise from molecules more directly, that qualia
might arise from—might *be*—the organised oscillations of molecules. Perhaps a particular quale arises when a particular group of
protein molecules over a sizeable region of the brain is somehow
"set going": oscillating in their particular, complicated way like
musical instruments contributing to a large-scale effect. There has
been some considerable interest in such ideas in recent years.[53]
They appeal to me, although it all sounds pretty crazy—even to
me. But I have learned not to be put off too easily by that (remember Isaac Newton and *his* crazy idea). So I ask you to pose the following question: "Let us suppose that qualia are made like this.
Might it help?"—and then we can get down to more mundane
matters.

For example, we then might ask: How, then, could qualia be
controlled? Somehow, it would seem, *this* particular collection of
protein instruments would have to be set going at just *this* moment
(over a second or so) with just *these* variations in intensity over *this*
period of time coinciding with *these*, *those*, and *the next* qualia of
different sorts: and all in a meaningful way so as to create the

experience of eating a boiled egg, or having an angry conversation, or quietly surveying a landscape on a windy October day.

We know of plenty of examples where the control systems of a machine (a car, a PC, a gas boiler) are easier to understand than how the machine really works. No doubt this will be true of quala-gens too. Is there a way in here?

# Chapter 8

# Switches

*it is a melancholy of mine own*

"How did you sleep?" is a question commonly put to overnight guests at the breakfast table. It is a futile question, of course, since the victim of this interrogation has to say "like a log" come what may. A much more interesting question would be "How did you wake up?" Or, if one really wanted to go into it, "Do you happen to know how your qualagens started up this morning—you know, the machinery in your brain for making feelings and sensations?" Now you might think that the only reasonable answer to such a question would be "Haven't a clue. May I have the marmalade please?"—but indeed there are clues.

The brain is remarkably conservative in its use of materials and in its basic biochemical techniques. The odds are that the qualagens will be made largely of protein molecules, as suggested in the last chapter like most of the clever stuff in cells. And the odds are that the way they are controlled will be conventional too: most probably chemically, that is to say they will be switched on or off, or adjusted, with molecular keys.

We have already come across such things. In chapter two I described the standard picture of how nerve cells communicate. An impulse generated in one neuron causes it to release neurotransmitter molecules, which may exert either a positive or a negative effect on other neurons it is connected to through synapses. Well, neurotransmitters are molecular keys. A typical neurotransmitter is a relatively small molecule which switches on a particular class of protein "receptors" in a receiving cell. Directly or indirectly, this causes a flow of ions into or out of the receiving cell and hence influences the chances that this cell will generate an impulse of its own. But I stress that this is only one example of chemical control for one very particular purpose. Molecular switching is the *usual* way the various activities of protein molecules are organised in the day-to-day running of all kinds of living cells: in, say, the control of the import, or export, or transport of small molecules; or in controlling the chemical transformations of molecules within cells, converting them from one kind to another.

It is true that we are largely in the dark when it comes to the question of how the qualia-making machinery actually works. But the suggestion that qualagenesis is most probably under chemical control fits in with much of what we know about the control of our conscious states, as we will be discussing in this chapter. Let us start further back in the control systems of consciousness, where there is more light.

The brain stem, if you recall, is an elaborate extension of the spinal cord. When seen under a microscope, much of it is a rather untidy looking network known as the *reticular formation*. It has miscellaneous nuclei, regions of grey matter "processors" rich in cell bodies, mixed in with the white matter "wiring." This brain region never shuts down and has long been seen as controlling the state of arousal of the brain as a whole, in particular the processes leading to consciousness. We wake up from there. This is the bit of the brain that first hears the alarm clock, so to speak, and then wakes up the rest of the brain, sending signals upward through the limbic area, thalamus, and on to the higher cortex.

More particularly, there are several nuclei in the brain stem containing the cell bodies of relatively small numbers of neurons that make neurotransmitters that are somewhat out of the ordinary. They are called "the monoamine neurotransmitters." For example, neurons whose cell bodies are in a pair of these places, the *ventral tegmental area* and the *substantia nigra*, make the monoamine transmitter *dopamine* and their axons terminals are to be found many centimetres away:

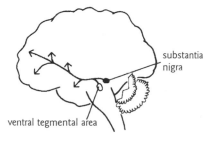

The axons of neurons are usually thought of as being like telegraph wires, devices for sending dot-dot messages to distant places. But a neuron can also be thought of as an exquisitely fine, and rather elaborate, hypodermic syringe with a branching needle, a needle of a thousand tips able to place minute amounts of a particular potent chemical in particular distant places. This description fits these brain stem neurons rather well. They have very thin axons and, as a consequence, rather low signal-transmission speeds. Although they place their magic juice with precision, they are not as precise as, say, the classical oversimplified "standard neuron" described in chapter two, where one axon terminal is firmly connected to only one place on one receiving cell. The dopamine released at the ends of the axons of these brain stem cells diffuse locally and are able to act on a number of nearby neurons, and perhaps on glial cells too.

To send your signals so slowly, and to be so unconcerned about exactly who you send them to—*that's* not the way you do the computing needed to keep complex activities on the rails. That's where you need to send the same rallying call to a fair-sized but distinctive group. That's Admiral Nelson raising his flag signal at the start of the battle of Trafalgar: "England expects every man to do his duty." There's nothing there, you notice, about how to set the sails of a ship in a rising sou'westerly, or what to do if your cannon jams. This was a signal, not a detailed instruction. But it was just the thing to prepare brains for action and switch them into "die-for-country" mode.

Like doling out the amphetamine? Well yes, more like that than you might think. Chemically, amphetamines are almost identical to dopamine, their molecules have a very similar shape, and they are thought to augment the natural go-to-it activity of dopamine signals in the brain. Perhaps they sometimes act as counterfeit keys taking the place of dopamine, acting as false neurotransmitters. This is often how psychoactive drugs work. But it appears that the principle action of amphetamines—and cocaine too—is more crass: They jam up the works. More precisely, these drugs block a set of protein pumps in the membranes of axon terminals whose job it is to clear dopamine from the synapses in readiness for the next nerve impulse. But the effect is similar: Too much dopamine is then left in these synapses, and too many go-to-it signals are thus reiterated to higher places—particularly to frontal areas of the brain where actions are formulated and carried out.[54]

There are other small brain stem nuclei containing neurons that have long thin axons branching off to even more varied and distant places. The few hundred neurons in the *locus ceruleus* supply another monoamine, *noradrenaline*, to almost everywhere in the brain, while the nearby *raphe nuclei* similarly supply *serotonin* sparingly but widely.

There is no doubt that these monoamime neurotransmitters are especially associated with qualia production. Drugs that "go for the qualia," that produce altered states of consciousness, often resemble these neurotransmitters or have been shown to enhance or modify the action of one or more of them. The effects of amphetamines and cocaine on dopamine levels are examples. Another is the effect of the antidepressant Prozac in increasing the activity of serotonin. But none of these effects is simple. A glance at my thumbnail map of the dopamine routes shows that several distinct brain areas are supplied. Furthermore there are several different kinds of "receptors" that are manipulated by dopamine keys, although for what exact purpose is not always clear. Serotonin has an even wider spread of destinations, and an even greater variety of receptors it acts on.

However it happens, moods are unquestionably affected by chemical means. The physicians and alchemists of antiquity recognised this when they used the word *humour* to cover both general mental states and bodily substances. The term "melancholia," for example, meant both mental depression and (literally) "black bile," an excess of which was thought to make you gloomy and withdrawn. We can agree with the general idea, but it would be going too far to say that our moods are identified with such substances. The monoamine neurotransmitters are not moods as such. They are only intermediaries, like the sailors raising Nelson's signal.

Moods often come over us quite slowly and seemingly in spite of ourselves—and perhaps from a kind of unconscious buildup originating in previous conscious thoughts and experiences. Any-

way, that is the what sad Jaques tells us in *As You Like It* using al-
chemical language:

> —but it is a melancholy of mine own, compounded of many
> simples, extracted from many objects, and, indeed the sundry
> contemplation of my travels which by often rumination, wraps
> me in a most humorous sadness.[55]

Of the psychoactive drugs, the major hallucinogens are the
most dramatic. As their name implies, hallucinogens make ordinary people see things that aren't there. Among other effects, these
drugs are *obviously* interfering with brain processes that organise the
qualia of our perceptions if not directly interfering with the qualia-
making machinery itself. Colours, for example, may become more
vivid than ever before, powerful feelings may appear inappropri-
ately, strange ideas and urges may develop . . .

Two of the major hallucinogens, mescaline and psilocin, are
molecular look-alikes of noradrenaline and serotonin, respectively.
LSD is a larger molecule, but part of it resembles serotonin. It acts
by either blocking or mimicking the action of serotonin in different
kinds of "serotonin receptors"—i.e., protein machines presumably
with somewhat different functions but which are all operated by
the serotonin key.[56]

The ergot mould has been known for centuries as a highly poi-
sonous contaminant of rye grain able to produce a whole range of
ghastly symptoms, including convulsions, in anyone unfortunate
enough to consume contaminated bread. In 1943, Albert Hoffman
was working with derivatives of this mould. He had gone home
early one day feeling restless and dizzy. He then found himself un-
usually but quite pleasantly intoxicated and soon hallucinating
wildly, seeing fantastic images and colours. He suspected that he
must have accidentally ingested one of the ergot derivatives, lyser-

gic acid diethylamide (LSD). Later, to test this idea, he cautiously took a tiny dose of the stuff (250 millionths of a gram) to see what would happen. It turned out to be a very large dose of what we now know to be an exceptionally powerful drug. This time it was not so pleasant, although he was lucky enough not to have a panic attack—perhaps he was too busy noting his experiences. Anyway, he had recovered completely by the next day.[57]

Many such experiences with LSD have been reported since. People talk of an initial euphoria followed by general distortions of perception. Objects are described as being seen as if through a kaleidoscope or as if reflected in curved mirrors. Sounds are said to be *seen*, and colours to be *heard*. Then in later stages there may be emotional swings, nice or nasty. From such varied effects, it seems clear that at least the switchgear for several different types of qualagen are being affected by LSD and other such drugs. By implication, the monoamine neurotransmitters they so often resemble are natural keys that control, among other things, our conscious states.

Chemical signals within living organisms often operate—and may be amplified—through a "cascade" of processes. Molecule A, let us say, causes a number of protein molecules B to make a lot of C which causes yet more of D finally to switch on E:

*initiating signal:* A $\rightarrow$ B $\rightarrow$ C $\rightarrow$ D $\rightarrow$ E $\rightarrow$ *does something*

In the brain, this sort of thing can happen via neuron activity too. The neurons that make dopamine neurotransmitters, for example, nudge neurons that make other neurotransmitters. So you can imagine A,B,C,D being in different parts of the brain.

It is a reasonable first guess that in the control of qualia production (as, say, with the control of muscle contraction), proteins are being switched by chemical signals all along the line; and the final effectors, the things which make something happen, are still proteins being switched by chemical signals in the usual way.

Let us say that E in the above chain stands for Effector as we recall the musical metaphor from the last chapter. Let us say that E is a relatively large set of qualagenic proteins ("instruments of a particular sort") making up a qualagen ("orchestra section") that can produce a particular sensation ("kind of music") when enough of them are caused to oscillate ("play") together, the intensity of sensation ("loudness") depending on the number of qualagenic proteins so activated. D is the chemical signal that switches on these proteins. So the presence and intensity of sensation depends indirectly on the presence and amount of D diffusing into the brain region(s) where the qualagenic proteins are located. This kind of picture, with many protein molecules being activated at the same time under diffusive chemical control, would account for the relatively long time lag of conscious phenomena; and perhaps particularly for the relatively slow onset of intense sensations such as severe pain or fear (as in the "I-didn't-have-time-to-be-scared" effect, frequently noted by heroes).

As I remarked earlier, the nervous system has in its neurons an excellent set of "micro syringes" for putting particular chemicals in more or less particular places. Perhaps this is how a molecule such as D in our example would be located. Then D molecules would be like neurotransmitters in the sense of oozing out of axon terminals. But they would be unlike the classic neurotransmitters described in chapter two in that they would not simply be helping to send on a message to other neurons: They would be *doing* something instead, making a quale by switching on qualagenic proteins.

What sorts of molecules might be best for these D molecules, these qualagen switches?

There is always a trade-off between speed and discrimination when it comes to chemical control molecules (molecular keys). If it's speed that's needed, a small molecule is best: a small amino acid perhaps, or a metal ion such as calcium. Small molecules diffuse quickly. But if it's discrimination you are after, then you have to go for something bigger so there is a larger variety of keys. There are only a few kinds of small molecules but immense numbers of big ones like protein molecules—immense numbers of ways in which the chains of amino acids that constitute proteins can be arranged. In between there are, for example, big amino acids or the monoamine neurotransmitters such as dopamine. As we have seen, the monoamines in particular are clearly implicated in the control of conscious states. But the numbers of such molecules are still quite limited. There are only a few that are actually used as keys in the brain. There is no way that each of the various sensations we are capable of having could have a monoamine key assigned to it. The key ring of possibilities is too small.

There might be two ways of dealing with this that fall in line with our speculations here. Monoamines such as dopamine or serotonin might act directly on qualagenic proteins to affect our states of consciousness, producing different qualia according to exactly where in the brain the monoamine is being released to act on the appropriate local qualagenic proteins. Alternatively, the monoamines are further back in the line of command, concerned more with general activation of, perhaps, not just conscious states but aspects of the Greater Self too. The second alternative seems more likely since monoamines are indeed associated with general activating systems of the brain stem as we have seen.

Peptides are another class of signal molecule. These are very small proteins, typically some 10 to 20 amino acids strung together, but there is still an immense number of possible peptides. A large number of so called *neuropeptides* have been found in the brain and it seems likely that most are yet to be discovered. The neuropeptides are often described as neuromodulators rather than neurotransmitters, and it is not always quite clear what they are for. But several neuropeptides have been associated specifically with states of consciousness—powerful and primitive qualia such as feelings of hunger, thirst, pain, and sexual desire.[58] And there would certainly be enough possible peptides to have one for each sensation we are capable of—if indeed this is what is needed—although here again *where* in the brain a neuropeptide is released might be important for qualagenesis.

Now recall our earlier discussion in chapter three of how we may often be conscious or unconscious in carrying out some activity; and how our consciousness of particulars may fade out as we become skilled in a task. Consciousness comes and goes, changes its focus; it almost seems like an optional side effect to be brought in as required.

Imagine standing in line at the bus stop wondering if that bus in the distance is the one you want. Someone vigorously and unexpectedly nudges you. Your big computer, your Greater Self, will have reacted appropriately before you become conscious of this potentially threatening situation, and it is no doubt already formulating some verbal response. Then you feel the nudge which has effects on both conscious and unconscious processes:

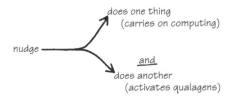

There is a divergent branching of cause and effect: one cause having two effects. But there are converging chains to think about too. Sometimes two or more things must happen for *one* other thing to happen: two signatures needed on the cheque, two keys to open the safe, or simply what we discussed in chapter six, a number of *contributing causes*—a bit of this and a bit of that in varying proportions, tuning effects or creating effects of different sorts. Or perhaps it is just that Bill needs *both* a nudge *and* a wink to be sure that he does what you want:

Now this matter of *converging* arrows is very much part of the problem of consciousness too—the other part as it were: how, in collaboration with Greater Self, Evanescent Self can have its say. So here now is the missing sketch from the beginning of the last chapter: another view of consciously walking across that hospital ward:

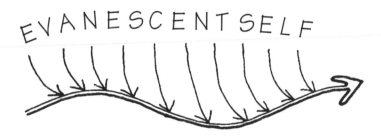

I don't know why it is, but many people who are prepared to accept that physical and chemical processes in the brain cause feelings and sensations—because really there is no way out—dig their heels in when it comes to suggesting that maybe it's the other way around too: that feelings and sensations can influence physical and chemical processes in the brain. It seems to me that if you swallow one, you might as well gulp down the other too. I suspect that both of these potions come from the same bottle, and that to fuss about the second spoonful is a bit like accepting that this mysterious thing the Earth's gravity keeps the Moon in orbit, but then saying that the idea of the Moon's gravity raising the tides twice a day is, well, just going *too* far.

Let's try taking both potions. Perhaps we could then combine the two big arrow diagrams from this and the previous chapter into something more explicit.

# Chapter *9*

# Arrows and Desires

*the cause is in my will*

$T$his is your brain:

$$B$$

And this is your brain as it was between 8.12 a.m. one morning and two seconds later:

$$B_{8.12.0} \longrightarrow B_{8.12.1} \longrightarrow B_{8.12.2}$$

This picture says that the state of everything in your brain at 8.12 was one of the factors determining the state of your brain one second later at 8.12.1, which in turn was among the causes of its state at 8.12.2.

There is nothing profound here, only the polite implication that your brain changes from moment to moment, but in a tolerably smooth way. Next we'll add some arrows representing events from the Outside World, OW for short, the world outside your brain

where there is another stream of existence, a world having effects
on your brain and in turn being affected by it:

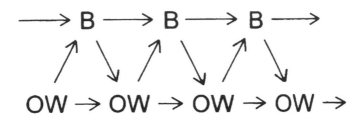

We can include in the OW↗B arrows *any* effect of the surroundings
on your brain. We are particularly interested in vast amounts of information that will from time to time stream in from your sense organs, although a change in blood chemistry, or being briskly hit on
the head with a mallet would be other examples of OW↗B arrows
at work. Likewise B↘OW refers especially but not exclusively to
muscles being ordered to contract or relax.

I forgot to say that you were still asleep in bed at 8.12 that
morning, with dreams coming and going, but you were not to remember any of them. Dreams like this are a particularly simple
form of consciousness, a partial form. They are close to being
"epiphenomena"—produced by the brain but having no further effects—and so we might represent these episodes of your dreaming
by intermittent streams of consciousness: C → C → C . . . being
maintained by some as yet undetermined brain activities (B↗C):

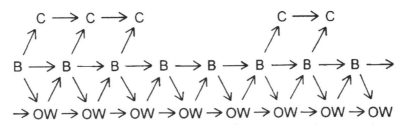

Actually the Bs have now changed their significance a little,
because we have now separated out part of the brain's activities for

special mention. B now means "everything about the brain except its qualia, its feelings and sensations, etc." B now refers especially but not exclusively to the unconscious neuro-world of Greater Self, while C is "consciousness" or, as I have been calling it, Evanescent Self.

A division of The Self into these two parts is justifiable, at least for the sake of discussion, as it is justifiable to talk about The Immune System, or The Circulation as distinctive parts of us. But there can always be border disputes in such cases. There will be questions about whether this cell or protein, or that water molecule, should or should not be seen as, say, part of The Immune System. For purposes of analysis, then, the division of The Self into two parts is not out of the ordinary, and it is not intended to mean that Evanescent Self has an altogether different domain of existence. The discussion in chapter four and the evolutionary argument of chapter five spoke against this. The unconscious and conscious aspects of the brain's activities belong to the same world all right, it's just that we don't know enough about that world to see how. This provides another reason, a pragmatic one, for distinguishing between the Greater and Evanescent Selves. For it is a distinction between aspects of the brain's activities which we understand in principle, and think we will come to understand wholly, and aspects of the brain's activities which we hardly understand at all. But of course that is just the way things seem to be now.

In the meantime the qualagens can be imagined to be on the border, lying along the B↗C arrows.

OK, so then you wake up and say "Oh my goodness, is that the time" yet still lie there, perhaps, feeling increasingly restless and uneasy until the point where your thoughts turn from uneasy to anxious to frantic. Spurred by these feelings, you start sending serious signals to parts of the world outside your brain, mainly to your legs, etc—and you jump out of bed. So we add some more arrows including a set of C ↘ B arrows so that all those guilty feelings might actually have had effects:

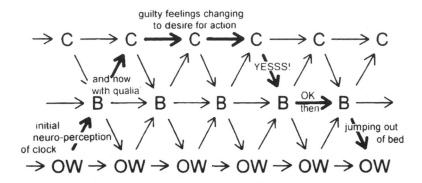

It follows from the evolutionary argument of chapter five that your feelings and sensations must be part of the mechanism of your behaviour as well as the unconscious mechanisms with which they are associated. So C ↘ B arrows are needed. But the C → C arrows are also implied by the evolutionary argument. In "going by the feel of it," there is a distinct set of conscious events which are taking place "up there" within the stream of consciousness and for which there is no complete set of unconscious "neural correlates" doing something exactly equivalent and just as well.

How long can one "hold things in one's consciousness"? Perhaps we can get some idea from the duration of short-term memory. This is the kind of example I am thinking of: You quickly read a 10-digit telephone number from the directory, and then immediately dial the whole number. If you realised half way through that you had hit a wrong button, you might be able to redial immediately without having to refer back to the directory. But if you had to re-dial, say, a minute later, you probably would not succeed.

In experimental trials where people are asked to memorise and recite long strings of digits, it is possible to get estimates of the duration of their short-term memory—their "working memory" as it is appropriately called. Clearly, if subjects were slyly writing notes on the palms of their hands this would spoil the experiments. The snag is that we commonly *do* "write things down" that we become conscious of, some of them anyway. Not on our hands, of course, but in our brains. We transfer some things that we have been conscious of to longer-term memories. That is a "re-remembering" kind of remembering, not a "holding-in-consciousness" kind of remembering. The real question centres on how long we can keep things in our heads without "writing them down" in any sense. It is about how evanescent the Evanescent Self is.[59]

Suppose a brain could be shut down, with all activity stopped, in a tenth of a second. Would consciousness persist for a second or two more? Some brutal and unsatisfactory "experiments" along these lines were in effect carried out in France around 1790 using a nasty brain-stopping device called a guillotine: There was some speculation at the time about whether the victims saw the inside of the basket their newly severed heads had fallen into. Who knows? (Well, I told you it was an unsatisfactory experiment.)

But the dreaming state may be of some help here just because it is an "incomplete" form of consciousness. When we are dreaming we are effectively paralysed. Our motor functions are safely out of action as the insane dramas of our dreams unfold. We may have desires in our dreaming, but we generally lack the means to do anything about them. We might guess that this is *because the C↘B arrows are out of action.*

This ties in with another well-known feature of the dreaming state: that we usually forget our dreams quickly, unless we make a special effort by writing them down, or talk about them, or go on thinking about them. That goes for lazier daydreams, too, from my experience in this area. Idle states of consciousness seem to fade quickly, lasting only a few seconds or perhaps tens of seconds, unless they are somehow recorded more permanently in the brain. Perhaps we can say that states of consciousness that fade quickly like this do so because there is no "writing down" in the longer-term memory areas of the brain, and that this is so for the same reason we are effectively paralysed during sleep: *because the* $C \searrow B$ *arrows are out of action.* It is tempting then to think of those few seconds that we hold onto our dreams as representing the limited durability of the $C \rightarrow C$ stream, a measure of the evanescence of the Evanescent Self.

Now consider a brain that is fully awake with the $C \searrow B$ arrows in action and memory working normally. Need it be a full-blown message or only a signal that travels in the $C \searrow B$ arrows to establish a memory? In other words, need there be a lot of information that passes, detailing everything that we are conscious of and which is to be remembered? Or can the message be something more like "*That's interesting: Keep a note of that please Mr. Secretary*"? After all, in a conscious perception it was the unconscious Greater Self that supplied all the details: The simplest message would be "Don't throw that bit away."

$B \nearrow C$ information flow is quite different in this respect: This has to be substantial to provide the information for a conscious perception, say, the perception of a landscape.

Now the point of all this back and forth between Greater and Evanescent Selves can only be that something can happen within the evanescent state that is helpful to the organism as a whole, and that was helpful for the control systems of other animals that came before us. It would not be to provide an element of randomness, because that is only too easily found in chaotic effects of many kinds: not to provide a more reliable channel of communication, because evanescence is hardly the stamp of reliability—surely the neuro-world would do better here; not to add to computing capacity or speed, because conscious capacity and speed are feeble in comparison to unconscious processes. No, the point is to bring in feelings and sensations, and the point of *that* could only be if such things can sometimes cause other things useful for the animal to happen in the world; things that might not otherwise have happened.

Imagine you are setting the table for dinner. You notice a fork slightly out of place. You feel it would look better a little further to the left. You lean across and make the appropriate adjustment. Here is the diagram again:

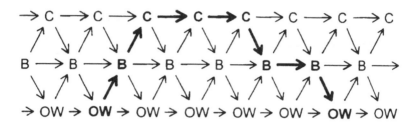

The outside world impinges on the brain, which recognises that something is wrong about the position of the fork (OW↗B)—more strictly Greater Self has (unconsciously) registered an error message. Unable to figure out exactly what it is, and/or what should be done about it, the Greater Self neuro-brain converts this information into a corresponding change in the pattern of qualia being maintained in the Evanescent Self, adding now an alerting "Hey-that-looks-odd!" quale (B↗C). After a couple of seconds (C → C → C) this Evanescent Self subsystem has felt its way to an answer, perhaps by adjusting the perceived qualia image to an imagined one of what should be—and then somehow conveying this back again to the neuro-brain (C↘B). Details of how to make the appropriate adjustment, and doing so, occupies the neuro-brain for another couple of seconds (B→B↘OW).

That our Evanescent Self indeed has physical and chemical effects is in line with the evolutionary argument of chapter five, but there are more direct indications too. There was the point touched on in chapter six that there are separate pathways for conscious and unconscious control of bodily movements. And, it has turned out, a monoamine neurotransmitter is very much part of a conscious pathway.

People who suffer from Parkinson's disease often find difficulty in controlling movements, more specifically in initiating movements and stopping them. ("My fingers won't do what I tell them."[60]) This has been shown to be associated with a failure of the substantia nigra to provide sufficient dopamine to the *striatum*—a complex central structure capping the thalamus, lying just above and forward of it, and known to be involved with an early stage of the brain's control of bodily movements.

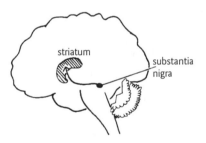

One of the great, largely forgotten tragedies of the 20th century was an epidemic of *Encephalitis lethargica*, "sleeping sickness," which started in 1916/17 and continued for a decade before disappearing as mysteriously as it had come. It left behind a million people dead and some three million human brains damaged beyond repair. Many people were to survive for decades. The story is recounted by Oliver Sacks in his wonderful, harrowing book *Awakenings*, and later was made into a film of the same title. The postviral symptoms of *Encephalitis lethargica* were complex, but the most characteristic symptoms were severe forms of Parkinsonism. There is a scene in the film where a patient, rather typically, is sitting inert, gazing into space, quite unresponsive to anything. But when he is thrown a ball, he suddenly and dramatically catches it, no problem. It looks like an example of that "duality of motor function" referred to in chapter six in which mechanisms for conscious actions may be inhibited, damaged, or eliminated while automatic actions are unimpaired. Sacks describes how one of his patients coped when, as sometimes happened, she "froze," unable to move:

> Thus, she had various ways of "defreezing" herself if she chanced to freeze in her walking: she would carry in one hand a supply of minute paper-balls of which she would now let one drop to the ground: its tiny whiteness immediately "incited" or "commanded" her to take a step, and thus allowed her to break loose from the freeze and resume her normal walking pattern.[61]

She could obey commands, and she could react to outside stimuli, even if covertly generated by herself, but in these frozen states was unable to exert a direct will of her own.

It thus seems that in the freezing of action that is characteristic of Parkinsonism a conscious intention is being blocked in its execution, but a broad intention, just to get things going. The details of what happened after that would be looked after by undamaged unconscious brain activities.

We had noted earlier that C↘B arrows might not be sending elaborate messages to Greater Self to consolidate a conscious short-term *memory* (because Greater Self would already have the information and would just have to be told to store it). Perhaps our C↘B arrows set *actions* going in a similarly austere fashion, in this case simply selecting from possible courses of action being continually worked out by Greater Self—that our Evanescent Self somehow provides the nudges for innumerable everyday actions, a faculty missing from Dr. Sacks's patient while she was in her "frozen" state.

Another of her frustrations, typical of her condition, was being unable to *stop* actions consciously, although she had devised ways of coping with this too, ways to "pace" herself as Sacks aptly puts it. When athletes use pacing techniques, they too are setting up external incentives, an implicit admission that their own inner will to succeed is not enough for the highest achievements their bodies are capable of. But at the same time they are demonstrating that they *do* have a conscious inner will of some sort that is distinct from these external incentives.

Our wants are a critical part of our consciousness. They are its *explanation*, because they are the instruments of its action. But I suspect the arrows of desire are blunter than we might think, and that we can only will the most general of intentions.

Moods are blunter still: but they are still instruments of action (or inaction). We know that our mood is likely to affect what we will do (or at least other people know this about us). Moods seem to be thrust upon us: We know only too well that it is not always in our direct conscious control to decide what mood to be in, although indirectly we may be able to manipulate it—with soothing (or rousing) music perhaps.

Moods, emotions, pain, and pleasure seem to be relatively simple elements of consciousness. This is in keeping with their broad, direct effects on our actions, and they seem to be primitive elements too, having their seats in the most ancient parts of the brain. The perceptual qualia, on the other hand, and the qualia of conscious thought are altogether more light-footed, more structured, more complicated—"higher activities" as we are inclined to say. Presumably, they call for particularly effective switchgear. Perhaps they run more closely with Greater Self?

# Chapter 10

# Humpty Dumpty

*in my mind's eye, Horatio*

$H$umpty Dumpty sat on a wall—and it was downhill from there. I have for you a somewhat similar tale: A tale of two eggshells.

When the doctor peers into your eyes with his little torch, he is looking at two bits of your brain which are out on stalks. The retina of the eye, curved like the inner membrane of an eggshell, is a neuron-based computer as well as being a thin, light-sensitive membrane. It has some readily identifiable camera equipment connected up to it: an iris diaphragm, an adjustable compound lens with a cap that blinks to keep the front of it clean. No shutter, though. The whole thing is more like a TV video camera, excellently mounted with great swiveling capacity.

Thus, sharp moving images of the world outside are projected onto these special bits of brain. Is that it then? Is that how we see the world—through movie pictures screened directly into our brain? That is certainly part of the story—the first part. But only a tiny part. From the point of view of easy understanding it's downhill from there. The shell-like retinal images fall to bits, as it were: they are analysed, decomposed, and their information scattered to different parts of the brain.

The analysis of information from the left and right halves of our visual picture of the world, for example, is soon farmed out to opposite sides of the brain with aspects of colour being handled in one set of places on each side, motion in another set. Early on, such things are dealt with in different layers of neurons within the thin fabric of the primary visual cortex. Information then passes to adjacent areas of cortex, and then to other areas altogether. Soon there are bits and pieces of activity all over the place in the brain, analysing information of different kinds, using it in different ways: computer activity all over the place, Greater Self activity. Yet in our *conscious* visual awareness, what we tend to think of as the end result of all this effort, there seems to be only one picture, in some sense made of sensations intricately and appropriately arranged, with its elements of colour, form, motion, etc. seamlessly united in a three-dimensional space. Humpty Dumpty had a great fall, but somehow or other he seems to have been put together again. In the next chapter we will be looking at some current speculations on this putting together. Meanwhile, how and why so much analysis?

Well, for a start the eye is not a window but a *transducer*. Like a doorbell, or a microphone, or a TV camera, it converts one thing into another. Press the doorbell and you set going another kind of signal: a flow of electricity in a wire. Speak into the telephone mouthpiece and it dutifully changes the pressure variations of sound into equivalent variations in an electric current to be remade into sound by the earpiece of some other telephone far away. Look at a TV camera and, while deadeyeing you back, it will be busy converting your picture into an immensely complicated signal, and sending it along a wire or on the back of a radio wave to be reconstructed, perhaps half a world away. Such transductions make the manipulation and transfer of information easier. Electric currents or radio waves can have patterns imposed on them; and they can be sent enormous distances without losing much of the intricacy of patterning they may have.

But why should *seeing* be complicated by transduction? We are our brains. We are in there behind our eyes, aren't we? Why can't we simply look at these little retinal screens from behind? Come to think of it, we are also behind our eardrums, which in their complex vibrating are a similar microcosm—in this case the sounds around us. And then our nose is close to two stalks of brain, "the olfactory lobes," which lie just above a thin bone high up inside the nose. Here there are cells that, in a sniff of air, can tell one kind of molecule from another and send signals about it via axons through the thin bone to those nosy pieces of brain nearby. And then, come to think of it, our brain is connected to a system of nerves throughout our body. Should we not include all of them as The Place of Me, so that we can presume to be out there in our fingertips or in any other part of our body? It's what it feels like, isn't it? And it's a reasonable view if we believe that our minds have the same kind of contact with the world that our bodies have. Not long ago there were many who would talk this way, as if our minds had direct contact with the world.

But it's not like that. As Shakespeare unaccountably never actually said:

there is transduction done

Yes, and again and again. By the time information from the outside world has penetrated a few millimetres into our bodies it has usually been converted from its original form as light or sound or pressure or heat or puffs of molecules . . . and always it is converted into signals of the same sort, *a common currency* let us call it, of electro–chemical activity of cells—*nerve impulses,* more or less. Your computer does much the same with the words, numbers, pictures, or sounds you put into it for processing. What your computer deals with, and stores internally, are as dull and colourless as those dot-dot-dot nerve impulses—in effect, strings of 0s and 1s. What can *that* sort of thing have to do with seeing or hearing?

The answer is calculation: things we think we "just see" or "just hear" or "just feel"—the direction a car is moving, that it is coming towards us; the meaning of a word, that we are in danger; the feel of a material, that it is silk. In hardly any such cases are our perceptions arrived at without some kinds of calculations, usually very elaborate calculations *based* on raw input data from sense organs but not simply to be identified with these data.

There are three parts to conscious perception:

Registering    Interpreting    Experiencing

Take vision again. The retinal images register something of what is going on outside. But that is not *seeing* in the everyday sense of the word, which includes interpreting and experiencing. If I say that I can see the cat, then in everyday language I mean that not only is the cat's image on my retinas, but that I know it is the cat, and furthermore I know it consciously.

Between the mere recording, which hardly anyone would say is seeing, and full-blown conscious visual awareness, which is seeing for sure, there is an intermediate subliminal level at which I would guess most of our "seeing" actually takes place. There is interpretation—understanding—in such unconscious "automatic pilot" activities as avoiding objects while walking across a room. So let us include all this in "seeing."

As a Reader's Home Project I would like you to think about how one might design a machine that could see in this sense, that is to say one that could interpret incoming visual information. What we would want is a machine that can see "a little bit," that could make an interpretation of *some* sort, and have an understanding at *some* level, however trivial. Consciousness does not have to come into it. I am not asking for an analogue of Evanescent Self. A simple model for the kind of unconscious perception made by Greater Self would be quite good enough. I suggest you read no further until you have a rough idea in your mind of how you might do this or why you think it would be impossible.

Such machines already exist. There are alarm systems which can see burglars by detecting movement in a room. Is "see" too strong? Of course they do not see the burglar as we might, with mask, lock picks, backpack, striped pullover, and all. And indeed such machines can easily make mistakes: they do not see at all well. But they can make an interpretation, and take action to prove it. They have a built-in "understanding" of this one aspect of intruders: that they move.

Of course, such understanding would have been put there by the designer of the alarm system. Is this cheating? I don't think so. Is there such a difference between this kind of thing and a kitten reacting to small moving objects? The kitten sees prey everywhere, you might say, and is usually mistaken. This aspect of the kitten's vision was also built in by a designer of sorts: Natural selection. The kitten is playing, enjoying itself no doubt, qualia abounding. But no doubt too, much of its response is unconscious, automatic, kittenish—Greater Kitten stuff.

So there are plenty of simple burglar alarms that can see, however slightly, in such an automatic Greater Self way. And we can readily imagine improved models, for example ones that would only react to objects of a certain general size. Or still more highly computerised security systems could become commonplace which, fed by data from a video camera, can recognise faces by making measurements on relative shapes and sizes, distances between noses, eyes, ears, etc, and perhaps taking colour into consideration too. Personal data could be programmed in so that the office boy or the chairman of the company is admitted or ignored.

In fact, work on such sophisticated "seeing" systems is well under way. Of course there is no need for consciousness in the kind of machinery being developed; no need indeed for the kind of global understanding which seems to come with our conscious awareness. A simple points system might do if the aim is mere recognition, for example, when enough characteristics of shape, size, colour, movement, and so forth match those of Bill Bloggs, then (most likely) it's him. Among the things necessary for any kind of seeing are computing streams, different kinds of them, to suit the different tasks involved in comparing two forms seen from different angles; or comparing two colours in different lighting conditions; or detecting and measuring movements of objects which may be close up or far away. You would want different sequences of operations for computing each of these things. They are different tasks requiring different kinds of calculation. The moving pictures in the video cameras used to collect the initial data would have to be the basis for several different *kinds* of analyses like this in order to begin to compete with the unconscious seeing that underlies most of our everyday activities.

For us, interpreting a scene is similar to what such surveillance devices do: We must first analyse it into different "features." It is what our retinas begin to do, and the rest of our brain continues to do relentlessly.

The paper-thin retinas contain several layers of cells. Near the back are the light sensitive cells, the "rods" and "cones." The rods are the most sensitive, able to detect light at very low levels. The cones are concentrated near the sharp-eyed centre of vision and, unlike the rods, are colour sensitive.

The rod and cone light detectors are the transducers of the visual system. Like other such transducers they are "half neurons." In this case, one end converts light into an electrochemical signal which the other end transmits to an overlying (transparent) network of neurons it is connected to. This network is the computer of the retina. It has its output cells, "ganglion cells," near the front, several kinds of which are interested in different aspects of the signals coming into the computer network.

The million or so ganglion cells of the eye have this many axons, or fibres, which bundle together to form the great output cable from the retina—the optic nerve. This makes its way back through the retina to the brain at "the blind spot" in each eye. The sketch is a "wiring diagram" showing what happens next (seen from below). The nerve fibres from the two retinas go to the pair of "lateral geniculate nuclei" (LGN) on the left and right edges of the thalamus, and these in turn send out a graceful fan of fibres to V1, the primary visual cortex at the back of the brain.

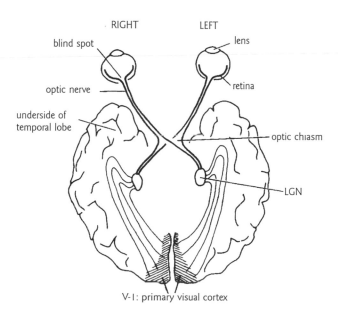

Note that the axons of the two optic nerves are rearranged at a junction box (the optic chiasm) near the centre so that axons originating from output cells in the left half of each eye end up on the left of the brain, while those from the right half of each eye go to the right of the brain. Because the lens inverts the image in the first place, the overall result is that the right half of your visual field picture is processed by the left half of your brain and *vice versa*. This is more or less in accordance with the brain's crazy arrangement of having parts of the body on the right side dealt with by the left brain and *vice versa*.

How do we see a colour? I am not asking the deep question about the colour qualagens, the question of how the colour qualia are manufactured by the brain; but rather the question of what kind of signals have to go to the brain for colour discrimination to be possible.

There are three kinds of cones in the retina which are informally designated "blue" "green" and "red," but all of them react to most kinds of light. Look at a red London bus and all three kinds of cones within your retinal images of the bus will react the "blue" ones very weakly, the "green" ones quite well, the "red" ones somewhat more strongly. So it is no use just saying that the "red" cones are doing their stuff in order to explain how further down the line your brain is able to come up with a red quale. A green tree on a bright day could easily be exciting the "red" cones more than a London bus on a dull day. The information the brain needs is relative. A comparison between all three cone types is necessary to get it right. So to know a colour needs calculations based on inputs from at least three adjacent cones (in fact many more are consulted), and this kind of calculation starts in the retinal computer.

Here is another puzzle: How can we recognise a politician's face in the few lines of a cartoonist's drawing? How can such simple drawings be so effective when they leave out so much? The broad answer is that delineating contrasts between light and dark is especially useful in differentiating one object from another—and the eye and brain know all about this. It's a long story and only partly understood, but it, too, starts in the retina, which is set up to take a particular interest in contrasts. It is almost as if the retina computer transmits something like a *drawing* of the retinal image to the brain—information that is somewhat pre-digested, somewhat interpreted, just as colour information is.

A ganglion cell receives inputs from some hundreds or thousands of rods and cones within a roughly circular central patch as well as from a ring immediately surrounding this patch. For one set of ganglion cells (about half of them), a small spot of light within the central patch will set the ganglion cell off. It will substantially increase its rate of firing. On the other hand, a spot of light in the surrounding ring will inhibit the signaling of the ganglion cell: It will fall to below its normal rate. These are so-called "on centre" ganglion cells. For the other class ("off centre"), there is an exactly opposite effect: A central light spot inhibits the signaling of the ganglion cell while light just outside the patch, in the surrounding ring, has a strongly activating effect. In each case a spot of light big enough to cover both the central patch and the surrounding ring has little effect either way. The retinal computer thus seems to be set up to inform the brain about details of regions of contrast; it takes little interest in areas which are uniform.

There are several kinds of ganglion cells which come in "on centre" and "off centre" versions. The most common are the P cells (P for "parvo," meaning small). They gather data originating from a comparatively small number of rods and cones and send information to the brain on the fine detail—high contrast and colour (subclasses of P cells deal with different kinds of colour contrast).

M cells on the other hand (M for "magno" meaning big) have a bigger catchment area for their data; Their acuity is poorer, but they are more sensitive to low light levels and better at discerning weakly contrasting features. They also have thicker axons which make them faster.

When they arrive at their staging posts in the right and left LGN of the thalamus, the axons from the retinal P cells and M cells go to different layers within these structures. There, ongoing signals remain differentiated and are sent to subtly different places in the primary visual cortex—to layers at different levels within the grey matter fabric. P and M now refer to distinct pathways or computing streams: P uniquely deals with colour but also helps with texture, shape, and spatial location; while the colourblind M stream seems to specialise in movement.

From different layers in the primary visual cortex V1, signals are directed to different striped zones of an adjacent area V2 as well as to more distant places. The "wiring diagram" has been deduced in some detail for our close relative the macaque monkey. More than thirty "maps" of the retinal image have often been found in distant parts of the brain. Such a "map" is a crude representation of the retinal image with points close together in the retina connected to points close together in the "map."

What can be the point of such multiple representations? Let's think of a seemingly simple situation: Imagine that you are on the shore staring out to sea keeping you eyes fixed on a sailing dinghy that you think might be in trouble. There is, let us say, a cat on the sea wall to the left in the direction of the lighthouse, and a dog to the right in the direction of the ice cream van. Then (even if you are not conscious of it) the cat would be analysed by the right brain and the dog by the left brain. If you are looking at the centre of the dinghy intently enough, even the right and left halves of it would be analysed by opposite sides of the brain. But in practice your eyes dart, fixating on different points of a scene, and your body and head move so that V1 on both sides of your brain will soon have raw data on most of the things around you. But with all this moving about—of somewhat different and partly overlapping left-and right-eye views—how does a single *stable* picture eventually emerge, with maybe a cat, a dog, a lighthouse, and an ice cream van included? In general, it seems that there would be no stable images in your head anywhere, not on the retinas, not in any of those "maps," nowhere.

Or imagine walking across a field of stubble. Everything on your retinas is moving in complicated ways. You do not even notice the apparent motion of the pieces of stubble as they move in relation to you, toward you and passing you on their various giddy trajectories. You are just aware of making your way across a more or less solid piece of ground.

Solid? Well yes. But where did you get that idea? And who puts the thirty and more pieces of Humpty together again—all those little maps? Indeed, who puts all the pieces of *two* wobbly, two-dimensional, mostly fuzzy, rather patchy, blind-spotted retinal Humpties together to make *one* stable, seamless, three-dimensional, qualia-laden view of the world?

Chapter *11*

# *Together again*

*a horse! a horse! my kingdom for a horse!*

You may recall that in the Humpty Dumpty affair:

All the king's horses and all the king's men
Couldn't put Humpty together again

It was perhaps a mistake to expect horses to be particularly helpful in repairing a broken egg. But can we do any better in explaining how an integrated conscious perception is arrived at?

Here are four possibilities:

1. A conscious perception becomes integrated as an unconscious perception which then becomes conscious; or
2. it is in becoming conscious that perception is integrated; or
3. there is a bit of each; or
4. neither: No integration happens or is necessary.

I go for either (1) or (3). In the last chapter we saw how fantastically complex the processes of visual perception are, even without touching on the question of how visual sensations are produced.

127

Earlier, we admired the efficacy of our unconscious systems which allow us, for example, to walk across a room without bumping into the furniture. To do something like that, you would seem to need a bit more than a stack of raw data, which suggests that at least *some* integration, some making-sense-of-the-data, has to take place at the unconscious level. Now surely Greater Self must lie behind Evanescent Self if only in providing perceptual information, controlling the patterning of the qualia that make up our inner perceptual world. We may surmise, then, that much of the integration of conscious experience has its origin in Greater Self.

None of this is to deny the possibility of an element of (2): that our conscious mind has a kind of integrity of its own, even if somewhat "evanescent," providing *some* of the integration of our conscious experience—so that (3) might be the correct answer to our question.

We can indeed get a feeling of there being a collusion of conscious and unconscious in a simple experiment in stereoscopic vision. By alternately opening our left and right eyes, we find that the images on our two retinas are different in detail, particularly for objects close up. These discrepancies provide information to the brain which helps it construct two kinds of representations of that part of the world that our eyes have registered. One of these consists of patterns of neuron voltages (part of Greater Self); the other is a conscious, 3-dimensional, quale-laden "moving picture" (part of Evanescent Self).

Above is a stereogram from which you can produce an illusion of depth as follows: Focus your eyes on a distant object and then, without refocusing, hold the two CABs at your usual reading distance. If you have managed to keep you eyes focused for distance, you should now see the double picture doubled, i.e. four images in a row. Now adjust your focus and/or the reading distance until the middle two CABs are superimposed. (Try concentrating on just one part of the two images, for example the top Bs and get these corresponding parts on top of each other.) You should now have three images the middle of which will become a stereo image. Keep examining it and over a few seconds its superimposed parts will fuse and a wonderful illusion of depth will develop, with ABC in alphabetical order. It is a very satisfying feeling if you can get this illusion, although there are some people (including one of the leading authorities on this phenomenon) whose brains cannot be fooled and who are never able to see it. Such readers will just have to take our word for it that there is a strong impression of there being underlying machinery at work which, over a period of a few seconds, is interpreting two, 2-dimensional images to make the illusion of one 3-dimensional "world." There must be quite complex calculations going on. The brain has to figure out which are the corresponding parts of the retinal images and make assessments of discrepancies between their positions so as to work out where to place objects, near or far. But whatever it is doing, you will notice that your brain is not only putting things in place. It is creating the place to put them. It is inventing a *sensation of space*.

Come to think of it, our brains are inventing sensations of space for us all the time, quite as much as they invent sensations of color. All these things are brain-made *qualia*.

So are our conscious perceptions of motion. There is a difference between being able to figure out that something is moving very slowly by observing how its position changes with time (as, say, with the hour hand of a clock), and having a conscious perception of motion: actually *seeing* that something is moving.

There is a well-known illusion which helps confirm that the sensation of motion is a brain-made quale—detachable as it were. If you stare at a waterfall for a few moments and then transfer your gaze to something that is stationary, say, to a nearby rock, then this rock will now seem, for quite a few seconds, to be drifting upwards.[62] In an experimental situation, a display consisting of concentric rings endlessly expanding outward produced a similar effect when the expansion was suddenly stopped. The stationary rings then *appeared* to be contracting. Using a brain scanner, it was possible to look for regional brain activity while subjects were watching the rings expand—and then determine what happened when they stopped and the illusion of contraction was "seen."[63]

Rather as V4 is a small specialist region for colour sensations, there is another small region V5 (or MT) lying at the junction of three lobes, (occipital, parietal, and temporal), which has to do with motion analysis, particularly the direction of motion. As expected, this region in the brains of onlookers "lit up," particularly when the expanding-rings display was moving. It remained active, however, when the actual motion had stopped, the activity decreasing concurrently with the fading illusion of reversed motion. Thus the brain activity correlated not with what was actually happening, but with what was being consciously perceived to be happening.

Then again, just as brain damage (presumably to V4) can produce a particularly profound kind of colourblindness in which colours cannot even be imagined, the qualia of motion may also be pathologically absent. There is a rare condition known as motion blindness resulting from damage to V5 whereby the patient does not have a conception of "motion as such."[64] Moving objects, things like trays, teacups, or people's hands, suddenly appear to the patient in new positions for no apparent reason. The patient has no immediate sense of that "quality of motion" attached to them, a quality so commonplace that we did not even know that we needed a perception of it.

Moving still further forward from the primary visual cortex V1, there are dozens of other areas of the brain concerned with visual information. Indeed there is a succession of places along what is often thought of as a "route" traversing the upper brain and ending in the frontal lobes.[65] In crude terms this route (really a pair of routes, left and right) seems to be concerned with *where*—that is to say with spatial aspects of perception, with how things are related to each other, how they are moving, as well as with guidance systems for moving among them. There is also a pair of low roads for other features of perception. These go via the temporal lobes to the frontal areas of the brain. The succession of places on this lower route are concerned with *what* rather than *where*: with aspects of identification and interpretation. Both routes have been studied in detail in the macaque monkey and they are presumed to be broadly similar in layout to that of man, its close cousin.

The increasingly "higher centres" are no longer just analysing and reacting to such elementary features as orientations of lines, shapes, and motions of patches of colour, and so on. They are increasingly dealing with those complex combinations of features found in real objects. They are also making connections with non-visual aspects of perception and thought. In addition, their activities become increasingly affected by what happens to have "caught the attention" of the brain. In their march from the primary visual area at the back of the brain to the frontal areas there seems to be more and more understanding possible. First they understand what is going on (in the parietal and temporal lobes) and then they determine what to do in the frontal areas. That is where the planning and action departments are and the switchgear for effecting movements of different kinds.

And these are not only processing pathways related to vision. They are memory places too, it seems, containing endless details of the *wheres* and *whats* that we discover in a lifetime of experience. When a subject is remembering a visual scene, or just imagining it in various ways, different parts of the these pathways may become more or less active. Or if a plan of action is being formulated in the frontal areas and needs a visual memory of some moving coloured object, then expect V4 and V5 to show activity as well.[66]

Maybe in all this we are beginning to see Humpty being put together again. The so-called higher centres in the brain are, after all, often concerned with making connections. Yet there cannot possibly be a physical centre for every possible combination of perceptions or thoughts. There cannot be a special place for every conceivable kind of face, say, or every possible sentence in every conceivable language. The brain must be able to put together elements in combinations that have never existed before—and quickly—calling for highly flexible communications between different brain regions.

At the most basic level of integrated perception, there is the question of how various "features" of an object that are being dealt with in different parts of the brain are perceived as belonging to the same object. This is called "the binding problem." There is no general agreement as to how this kind of thing happens, but "brain waves"—oscillations in brain voltages—look as if they may play an important part.

One of the first modern methods of studying brain function was by analysing brain waves—electroencephalograph (EEG) patterns. Electrodes attached to different places on the scalp will record rapid, oscillating voltage changes that reflect changes in the activity of the cerebral cortex immediately underneath. Shut your eyes and a voltage pulse of about 10 beats per second will be widely detected. This seems to be a kind of standby activity that masses of neurons switch to when they are not doing anything else. When your eyes are open, this so-called α-rhythm is suppressed to a large extent and more complex patterns develop. Among these are faster β- and γ-frequencies (15–30 pulses per second and about 40 pulses per second, respectively) and these appear to have something to do with the integration of brain activities. Anyway, they tend to become synchronised in different areas carrying out related tasks.[67]

In a recent study,[68] cats were taught to respond to images appearing on a screen by pressing appropriate levers so as to get a reward, maybe a blob of cream. Meanwhile cell oscillations in precise brain regions were monitored. While engrossed in their task, there was a tight synchrony in the oscillations of cells in different brain areas, between primary visual and parietal areas on the one hand, and parietal and frontal on the other—appropriate areas for looking intently at an image, choosing and pressing a lever, and perhaps thinking about cream. But this synchrony between oscillations relaxed when the task was over and the reward was produced.

Of course, this can only be the beginning. If we are really to put Humpty together again, it is not enough just to *bind* all the pieces together. They have to be arranged correctly; and for our perceptions, thoughts, and action plans, this must call for much more subtle and detailed forms of communication than just: "This goes with that."

If you had to design a means of connecting distant parts of the brain, you would want not only long "telephone wires" (as are clearly visible in, say, the mass of white matter axons lying within

the cerebral hemispheres), and not only a simple drumbeat synchronisation. You would want a system for selecting which of an as-
tronomical number of possible connections should be established.
There would have to be some kind of "dialing up" and/or "tuning
in" system with, perhaps, identifying codes.

Just look at this microscope drawing made by Ramon y Cajal
about a hundred years ago of a *very small proportion* of the neurons
in one microscopic part of a cortex.[69]

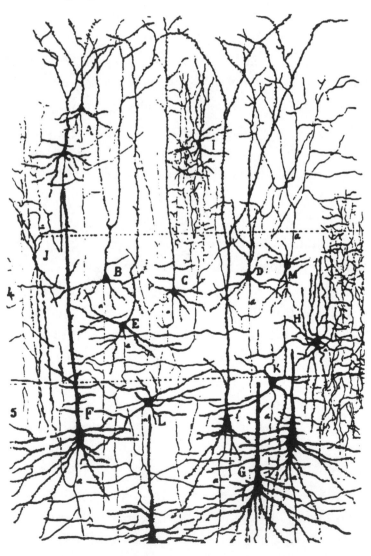

Imagine trying to send a message along telephone wires like that. "We're trying to connect you" . . ., "Hello" . . ., "We're trying to connect you" . . ., "We're trying to connect you" . . ., "Hello, is anyone there?" . . .

How can such a jumble be so organised? Recall the simplest idea about how neurons work, that they send signals to each other and vote "yes" or "no"about whether a receiving cell should fire a signal off to *its* recipients or not. Extra information over and above a simple yes or no is provided in such an arrangement by the rate at which these signals are being sent—the number of yesses or noes per second. It may represent, if nothing else, a level of enthusiasm. Or firing rates might be used for "binding," with neurons firing at the same rate all coming into synchrony and thus affecting each other more strongly. But really, any such "rate code" would still be a very limited way of coordinating the activities of rapidly changing groups of neurons in distant parts of the brain, the kind of frenetic activity that goes on in the most trivial perception or thought.

Another idea is that there are "temporal codes" in the signals that are sent between distant parts of this fabulous net. This is something more like African drum music with bursts of well-patterned activity which have more complex meanings.[70] Or we might liken it to a supermarket bar code which, when scanned, generates a succession of pulses which might mean: "Put 57 p on the customer's bill, and note that there is now one less small package of Uncle Andy's Low-Calorie Semolina on the shelves."

No doubt "rate codes" are part of the story, and perhaps the synchronisation of simple oscillations is also part of the way that different aspects of brain activity are locked together. Complex temporal codes are more speculative, but neurons are far more complex and sophisticated than is suggested by the standard "voting" story of how they work. In support of the idea, neurons—and especially groups of neurons— should in theory be quite capable of generating and recognising such temporally coded information.[71]

However it turns out, it seems likely that *some* kind of "combinatorial" codes, using elements that can be arranged in millions or trillions of different ways, must be involved in coordinating brain activity—which in real life has to cope with and choose from immense numbers of possible permutations.[72] One gets an idea of how efficient combinatorial codes can be for this kind of thing by thinking about a familiar example of such a code: a telephone number. The entire population of the Earth could, in principle, be assigned a *different* nine-digit number. There must be a system at least as discriminating as this, I think, to unjumble the activity of the cortex.

How does the brain work as an unconscious controller, and how does it create our conscious awareness? These two questions have been there or thereabouts over the last five chapters. The first is difficult still, but mainly because it is so immensely complicated. There is at least a sense, now, of being on the right road. And the second question may seem less baffling through its evident connection to the first. But there is another question that has become *more* baffling the more we understand the brain as a kind of neuronal-control computer. *Why* does it create our conscious awareness? That question, too, has been there or thereabouts almost from the start of the book. It is time to confront it more explicitly.

# 4

# The Secret Agent

## devices and desires

*O*ur perceptions are representations of the world. They are models of the world, but not models that anyone looks at. Not even our Self looks at them exactly: They are part of our Self. They are made to be working models, key components of our control systems, agents that act upon the world. The Body Image, for example, both conscious and unconscious, is a representation of the most personal part of our world, and it too is more than just a passive image. It is the part of our Self through which we most immediately act. And all of it, unconscious and conscious, is brain-made.

In Part II we had concluded that our conscious mind is part of the physical world all right, but not as that world is yet understood. Here, in Part IV, we will look at the conscious mind more from the point of view of an engineer than a scientist, seeing it as just another clever component of the body's control machinery. Perhaps we are in the position of someone trying to understand how a radio set works, but who as yet knows nothing about radio waves. That's what it looked like in Part II. But such a person might discover radio waves—or at least see the outline of the physical entity that would be needed to explain the seemingly magical radio set.

Perhaps we can see an outline here—the outline of a story. The underlying stuff of the world is energy. It may be woven into atoms and molecules; and then, on a larger scale, into matter as we normally understand it. That is one kind of fabric, a wonderful fabric—the fabric of life, no less. But the device we seek to understand weaves another kind of fabric out of the same energy: an evanescent fabric of feeling and sensation from which the agent of our conscious mind is woven.

# Chapter 12

# Why a phenomenal world?

*it is to be all made of fantasy*

The auditory cortex is not a particularly noisy place. Perhaps being quite near to the ear holes it is a trifle more so than most, but the real work of an auditory cortex is silent. The transducers in the inner ears turns that intense sound of Louis Armstrong's trumpet into silent, complicated nerve-impulse patterns to be silently sent on to the brain stem, thalamus, and cortex. Much the same can be said of parts of the brain concerned with the sense of smell: They do not exude the odours, the sensations of which they are creating within them—at least I don't think so.

The visual brain is another story, some might say. Here there is something more literal: actual images of the world—first on the retinas (although upside down and twice) and then again in stacks of somewhat similar "maps" in the thalamus; and then in more and more such "maps" in the cortex. But really it's the same story. Once the retinas have converted light into coded patterns and sent them on their way along the optic nerves deep into the brain, why then, all is dark. Not that it matters. Just as nobody is in there listening or sniffing, nobody is trying to look at maps either. To have various versions of the retinal image in different places most probably has some mundane explanation—saving on long wires perhaps.[73] Whatever the explanation, it is a different world once we are beyond the sense organs and into the brain: a world of computation, of data handling in dense networks of neurons, a world of codes and symbols. Not the real world at all but an abstract representation of it.

Then, of course, there is the world of our conscious experience, *the phenomenal world*. This seems at least as different from *the neuro-world* of voltage codes as this neuro-world is from *the real world*.

Colours, sounds, textures, hungers, pains, wonders, desires, interests, hopes, fears . . . The brain makes such elements of our phenomenal world as a distinct additional enterprise on its part. It is another step. As touched on in chapter four it does not look like a step back again toward reality from the world of codes and nerve impulses. It looks more like a step further from reality. Qualia are the stuff of phenomenal experience, and few would suppose that any of *them* are really "out there." Qualia are made by the brain— one might almost say invented by it.

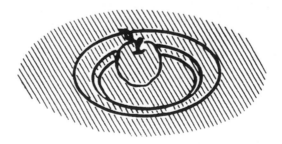

How do we see an apple on a plate? An apple is mostly water—some ten trillion trillion molecules of it—together with other more complex molecules: glucose, polysaccharides, proteins, lipids, DNA, and so on, mainly within billions of cells joined together to make different tissues . . . Oh well, let's forget about trying to say what an apple is. And as for the plate, I'm afraid I will have to pass that by too. I will forgo describing to you the wonderful crystalline silicate structures into which many of its atoms are arranged, or to get you to imagine what that surface scratch might look like when explored with an electron microscope at a magnification of a million—where a millimetre would come to look like a kilometre if only the field of view could be big enough. It would look stunning I should think, a veritable canyon. But all this we must lay aside. Ridiculous really. Imagine claiming that we can see an apple and a plate when we can only take in *so* little of them.

The shading over the above picture is a kind of apology for such a scant representation of reality—and for an admitted ignorance as to what on Earth *reality* is. Take the picture to be what computer people would call an icon, a trivial brain-friendly reminder of a complex topic; and then take the shading to mean "Well of course you realise that it is not really like this." Perhaps it would have been better to have just written "APPLE ON PLATE," but that would have led to other ambiguities . . .

OK then. So you are looking at these physical objects. The first thing that happens, according to the accounts in the last two chapters is registration (R): the production of two upside-down retinal images. Then there is the first transduction ($T_1$) to dynamic voltage patterns, in neurons mainly. This transduction is wholly understandable in computing terms—briefly, it is to be able to make calculations (C). More particularly, it is to analyse the images, interpret them, and make connections with nonvisual inputs, and with memories, and so on. And then such detailed "automatic pilot" adjustments that are needed, say, for bike riding can be worked out promptly—and executed. Indeed, all the major areas of the brain are likely to get involved in some such fashion in most of the things we do:

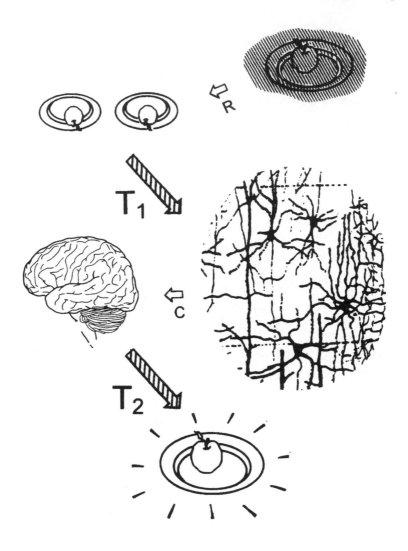

Then there is the second transduction ($T_2$) to a phenomenal image of an apple on a plate. It is an image set in the phenomenal world of our conscious being.

No more shading now because although the drawing is a grossly inadequate representation of this phenomenal apple and plate, it is not so *utterly absurdly* grossly inadequate as the other drawings.

Indeed, I could go on to tell you a bit more about the whole phenomenal experience of an apple on a plate, which might have included what these objects felt like in the hand; their weights, and coolness, and their surface feel; what it felt like placing the apple so that it stayed put on the plate, as well as its inferred deliciousness from longer memories of apples past; and how it nearly came to look like this:

From registration to the phenomenal image—such a thin and cautious description of perception is certainly an oversimplification and may be plain wrong. It is at best oversimplifying by having the journey from the real world to the phenomenal world written in a single path like that. There is much branching along the way. Furthermore, $T_2$ is not just a final link in a chain of events—at least not for normal forms of perception where what we consciously perceive at one moment can affect what we will perceive next. What does seem clear, though, is that $T_1$ and $T_2$ are distinct, and that for any particular item of perception $T_1$ comes before $T_2$.

But why go to all this trouble? Why are we conscious at all?

So far we have seen our phenomenal experience as one of two *representations* that the brain makes of the real world. So we should start by asking what the point is of having representations at all, or *models* as we may call them. Of course by the word *model* I do not mean people on catwalks showing off strange clothes, nor do I mean paragons of perfection. I am using the word in a science/engineering sense, meaning things like model aeroplanes, or maps, or physical representations of atoms and molecules, or computer models for predicting the weather.

Any model is an analogy of one kind or another, and like all analogies there is no problem if it is radically different in many respect from the thing it otherwise purports to resemble. In particular, the materials from which models are made may be different, although by no means irrelevant to the model's usefulness. A road map is better if it is not made of asphalt, but it has some rather literal elements in it all the same. For example, a long way on the map is a long way in the landscape. But then the representation of the church as a cross is not much like the real thing and is absurdly out of scale, while the route numbers written along the road lines of the map are not placed like that in the landscape. None of this is carelessness on the part of the cartographer. Maps may be all the more useful for suitably restricting their similarities to Nature, and in being selective in what they include. Some of the most useful maps, for example the map of the London underground, are the most abstract, but excellent at displaying what you really want to know to get from Euston to Heathrow.

When we say that the brain makes representations of the real world in terms of nerve impulses and so on, or that our computer does something similar, we are talking of "maps" that are a good bit more abstract than those of London's underground.

"Abstract" might not be the word that immediately comes to mind when talking about our vivid inner world of consciousness. We believe in its literal reality. But I would guess that this representation too is made more for convenience than literal realism. Perhaps my main reason for thinking this is that a model doesn't *have* to be literal.

Kenneth Craik was perhaps the first to suggest in the early 1940s that the brain makes models of the external world, but his were not passive models like transit maps. They were working models:

> I see no reason to suppose that the processes of reasoning are fundamentally different from the mechanism of physical nature. On our model theory, neural or other mechanisms can imitate or parallel the behaviour or interactions of physical objects and so supply us with information on physical processes which are not directly available to us.[74]

The kind of model Craik had in mind was a mechanical or electrical mechanism with its own workings being analogous to the way the world works.

Every child knows that there are two kinds of model aeroplanes: the boring ones you put on a shelf and try to admire, and the fun ones that fly. (And who cares if they do not look much like a real aeroplane?) A Craik model "flies."

The snag, you see, in trying to use mere passive representations to explain human perception is that passive models are made to be looked at. By *whom* we may well ask—and soon we have the crazy idea that there must be an inner perceiver for the brain's representations of reality, implying further representations of the representations in this perceiver, and so on and on in an infinite regress. The way to avoid this kind of runaway explanation is not to panic. In the first place, perceptual representations are not just look-alike images, they are radical transformations. In the second place, they are Craik models. They are part of a self-acting control system. They are not shelf models. They can fly. In the third place, and most crucially, these conscious and unconscious representations of reality are not models which "we" perceive or use as if there were some other "we" in there. I hope this doesn't come as a shock, dear reader, but they are two parts of what you and I *are*.

This bald assertion is the instant cure for *polyhomunculophobia*, the condition just referred to (and which otherwise has a rather poor prognosis) in which the patient, usually a philosopher, fears an endless nesting of homunculi—of humanoid perceivers within perceivers. Only an identification with *self* at some stage can put the stopper on this kind of infinite regress.[75]

But then we might ask again: Why this *phenomenal* world? "Maps" do not need to be conscious, they need have no feelings and emotions, and neither do Craik models. (Craik himself was quite clear on the matter: He was not intending to explain consciousness.) On the face of it, neuro-world representations seem perfectly adequate. And to have two different control systems might seem somewhat extravagant—and not to put too fine a point on it, asking for trouble.

We discussed, particularly in chapters three & eleven, how qualia may either be present or absent for some activity, both in normal people and in people experiencing pathological states; and it seems clear that there are specific brain activities devoted to producing qualia of different sorts—as we have seen, particular brain locations have been implicated in a number of cases. Phenomenal representations must be useful, presumably as part of the brain's overall control system, or the machinery required to produce them would never have evolved. But perhaps instead, or in addition, there is another answer to the *why* question. It may have been *easier* to evolve a dual system than to do everything with a purely neuro-electrochemical system.

It was part of Craik's idea that the brain contained "a physical working model which works the same way as the process it parallels:" That is to say the way the model is made, and the kind of stuff it is made of, will be important in creating the essential abstract correspondence between the model and the reality it is supposed to represent. No model is going to be ideal for everything; the analogy is going to break down in places.

So perhaps there is at least a very general answer here to the question "Why a phenomenal world?" if asked in the sense of: "Why a second kind of representation that is so completely different?" Perhaps it is because although the neuro-world computer does a wonderful job and maybe *could* perform all the functions we associate with conscious activities, it would nevertheless be surprising if just one kind of model was best for every purpose; and no great surprise if another kind of model, without the same neuro-electrochemical basis, would be more efficient or easier to evolve for some purposes.

But what are the particular virtues of consciousness? We had decided earlier that bike riding is mostly a Greater Self activity, and it is true that the detailed calculations required are almost entirely executed at the unconscious level. But consciousness plays a role all the same. If you decided to take the old bike to the store this morning, it was most probably a conscious decision. And perhaps you became consciously aware that there was something wrong with the front wheel and you had to pump up the tyre, and that the saddle was a bit too high and needed to be dealt with too. Maybe you were rather wobbly as you started off on your journey having not ridden the old bike for a bit, but you would soon get the feel of it again. That is it say Greater Self would have stopped sending you warning signals and be communicating, in effect, that everything was now OK at the complicated deeper levels of what you were doing. Your Evanescent Self could then concentrate its limited capacity on not losing the way. Your Greater Self would soon let your Evanescent Self know, with an appropriate feeling, if conscious attention to deeper matters was again needed. How else to communicate than by drafting the request in terms of the only thing the boss understands—qualia?

All qualia have a *quantity* aspect. In spite of their name, qualia are not just qualities. You can be more or less hungry. Just how hungry might tip the balance one way or another about that lunchtime talk you were thinking you should attend. Our sense of space and our sense of motion also include strong metric elements. We not only feel the sensations of distance, of movement toward us and so on, but these, too, come as *more* or *less*, more or less distance or movement in different ways.[76]

We use such sensations for making judgements—"calculations" we sometimes call them—about whether there is room to overtake the car ahead, or where to aim to hit a moving target. We make these judgements by feel, insofar as Greater Self has called up a conscious involvement—not by arithmetic.

The television commentator assures us of what we can already see—that the snooker stroke was beautifully calculated—but gives us no figures on the particular velocities and rebound angles which the player would not have known exactly either, but which had been required to execute his plan. Consciously, he had gone by the feel of it. Unconsciously his neuro-world model had helped him, calculate how hard to hit the cue ball so that it would come to rest under the cushion behind the pink, and then presumably informed the Evanescent Snooker Player what it should feel like to be about to do just that.

Snooker players, darts players, bike riders, any of us, any time we are awake, feel the way to decisions about how hard, how fast, and exactly in what direction we should go.

Let us now untie our question a little further. What are the particular virtues of having a representation of the world that is in some sense "made of qualia"? Here let me remind you of my assertion in chapter three that the defining feature of conscious awareness, conscious thought, conscious anything, is that in all such cases "the dish is served with qualia"—so that consciousness can be said to be "an arrangement of qualia." Before trying to see why it might be a good idea to have a qualia representation of the world, let me return for a moment for my reasons for describing consciousness like this.

Surely, it might be said, there is more to a conscious perception than just the added feelings and sensations, however broadly we may try to define such things: The dish may be being served with qualia, but that is just the sauce. And then surely, it might be said, a perception contains plain facts; and thought is guided (certainly *should* be guided!) by logic as well as feelings. Surely facts and logic are the main elements that determine what we think about and the rules we use to do so.

Let me explain why I would agree with much of that argument, yet still insist that consciousness consists of qualia and, in an important sense, nothing else. This is the sense in which maps are nothing else than paper and coloured inks, or a motor car engine is nothing else than various metals, etc. And it is an important sense. It is about the stuff they are made of.[77] No matter how interesting and complex the map is, no matter how much people may insist (correctly) that the arrangement of dyes on the paper is critical, if the paper disintegrates or the inks fade there will be no map left; or if the metals all corrode to nothing, no motor car engine; or if the Cheshire cat vanishes, there will be no smile left behind.

Indeed qualia, particularly visual qualia and the qualia of thought, are usually combined in complex ways—usually highly structured. But that is not to say that if the associated qualia fade, *consciousness* will remain. Perceptions may remain, thoughts may remain, and volitions may remain to control actions, but now only unconsciously. Unlike ordinary maps or Cheshire cats the critical structuring—the information, the programmes, whatever we want to call them—can still be there so long as the appropriate underlying structures and activities of Greater Self are still there. There need be no sharp switchoff. It might be difficult for an onlooker to discern when Sylvia became oblivious to what people might be thinking about her hat. "Conscious perception," or "conscious thought," or "conscious actions" are always mainly unconscious processes which become entirely so when the qualia fade.

So finally in summary, here are four answers I would give to the question of this chapter in its now modified form: What particular use is a qualia model of the world that it should have evolved in our ancestors?

I think the first qualia to have emerged would have been coercive sensations associated with the deep, ancient limbic structures of the brain: pleasures, pains, itches, nasty smells . . . Of course, we don't yet understand *how* such things arise in brains, but there is no problem of *why* given that coercive sensations arose, they caught on.

It is more difficult to see in evolutionary terms why we have so many qualia that are noncoercive, or nearly so: neutral sensations of colour, form, and so on, which are neither particularly nice nor nasty, but more purely perceptual—like *most* of the sensations that make up our complicated moment-by-moment consciousness. Perhaps it is because qualia, for all their rich variety, are alike enough in their physical nature to be combinable into a single, useful, simplified "world"; like oil paints, perhaps, with their rich range of colours that can nevertheless be handled together in the same oil medium (although it must be said that a state of consciousness is a good bit more versatile, active—and fugitive—than an oil painting).

Now why should thought ever have become conscious? Conscious thinking processes are often about perceptions, but they are suffused with more primitive and coercive feelings too. We judge whether something we are thinking about is true or false, we decide which direction to pursue our thought, we test "taste" difficult ideas, we feel the quality. We "ponder"; in other words, "weigh things up", as if deciding by muscle feel. Well, not muscle feel, but something close, and something we might surmise that had evolved from the bodily sensations which may still accompany them.

Whatever the virtues of a qualia-world there is a quick answer to our original *why*. It is that an animal's phenomenal world, however simple or sophisticated, is a device for making decisions. For us, it is especially useful for making difficult decisions; not necessarily the really important decisions of life and death, like deciding to look both ways before crossing the road or jumping in to try and save your child from drowning. These may happen with little conscious involvement, the actions programmed from habit or genes. We all know the difficult decisions: It is choosing the best candidate for a job, or the tie to wear for the reunion dinner; it is where there are lots of different things to take into account, and neither you nor your genes have quite been there before.

But how on Earth could anything as airy as "an arrangement of qualia" have any influence on the brain's workings?

# Chapter 13

# A working image

*in action how like an angel!*

*I* think it was unfair of people to criticise Mrs. McGinty because she wouldn't lift a finger to help her sister. Why should she perform a miracle for a woman she doesn't particularly like? And isn't it a kind of miracle that we can do this?—lift a finger, that is.

Try it. Hold your left hand in front of you so that you can see all five fingers, relaxed and slightly bent. Then think of all the things you can do with them. You can move any one you want, up or down, side to side, straighten it, crook it—there is all this choice, too much perhaps. But let's say you come to your decision. You decide against lifting a finger. You crook your little finger instead. How did all this happen?

Such body-moving skills are part of what it is to be an animal, we might even say its essence; and it might seem odd that we are so ignorant about them. For sure, we know something of what was going on. We know that nerves relayed messages from your brain to the appropriate little-finger-moving muscles, although perhaps like

me you are not altogether clear about which muscles had to contract and relax to do the trick, or indeed even where they are—still
less which sets of brain cells were working away at whatever it is
they work away at. We have so little conscious knowledge of how
we consciously act—almost effortlessly, in action so like an angel,
we just do it.

So if we were to ask Sylvia *why* she tapped the key for middle
C with her third finger, we would likely get a more or less satisfactory answer. But to the question of *how* she did it, we would likely
have to be satisfied with a "Well, I just did." Yet we might have
thought that if Evanescent Sylvia is an agent at all, if it is able to
influence her brain's activities in any way according to how it feels,
then there should be some sense in which this agent must "know"
how to prod Sylvia's brain appropriately. "Come on brain, do what's
needed to get that finger onto the keyboard. Yes, that's right, middle C. Now a quick tap." No one is conscious of how that sort of
message gets across. At the brain end, we are only conscious of a desire, or an intention, and perhaps something of the effort that will
be needed. At the finger end, we may sometimes be conscious of
what we like to call detailed movements, but even if you are a
watchmaker these are coarse indeed in comparison to the molecular mechanisms underlying nerve and muscle action. Of course, a
pianist like Sylvia will not be consciously aware of what all her fingers are doing in performing a well-practised piece: but even when
she was learning it and thinking about which fingers to use for
which notes, her explicit knowledge of what she was doing was exceedingly limited.

I am reminded here of our perceptual ignorance of apples and
plates as discussed near the start of the last chapter, and how we had
to own up to knowing so little about *that*. Just as we could see that
our level of perception is not down among the microscopic and mo-

lecular details of the things around us, so it is clear that the level at which we consciously act is not in the details either—not down among the neurons and nerve fibres, or even the tendons and muscles. What we consciously imagine ourselves to be is an absurdly curtailed version of all we physically are. We imagine ourselves rather as a painter or sculptor might represent us, showing little of what is beneath the skin. We do not need this fine-grained knowledge in our consciousness, just as a car driver does not have to know all the minute intermediate consequences of turning the steering wheel or putting her foot on the accelerator. In most of our actions we have no need for conscious knowledge of whatever it is that happens between a broad intention and a broad result.

It would seem that like the secret agents of fiction and fact, the agent of the conscious mind operates on a "need to know" basis—it does not have access to more information than is required to do its job.

Yet we all have an informal anatomical knowledge: our *body image*. This means a good bit more than just the sight of parts of our bodies, or our visual imagining of it. It is more a "feel thing" than a "see thing" when we become conscious of it. We know without looking whether our left little finger is straight or crooked, and we can imagine moving it, and we know what it feels like if we do.

Our body image has too light a name. It is no incidental thing. It is a crucial part of us, deeply built-in, part of our perceptual world. As such, it would have been invented for the most part in the course of evolution, as surely as our kidneys or our immune system or any other part of us. It's largely with us at birth, although finished in babyhood and honed ever since.

What we might be inclined to describe as "our finger" is, in truth, our perception of our finger. It is part of our body image based on touch and muscle feel and consolidated by all the senses. What you saw when you looked at your hand in the first place was part of a body image made in your brain. Like all forms of perception, both conscious or unconscious, it is part of you, but at the same time a representation of part of the world.

Like everything else in the real world, then, we have no direct knowledge of our body—only of a brain-made version of it. It is hard to believe this, and more than a little scary. But there is good reason for attaching a "guaranteed brain-made" label to our body image. People who have had a limb amputated usually report that they can still feel it. They have "a phantom limb" as it is called, and, unhappily, sometimes "phantom pain" in it as well. This is a false conscious awareness. But there is a false unconscious awareness too: patients will often try to use their missing limbs in habitual actions, such as reaching for a glass of water—the kind of things we do "without thinking."[78]

Even more bizarre are opposite cases, where a limb which *is* there is thought to be missing. This is a sign of brain damage (usually of the right parietal lobe of the cortex). A patient may lose the sense of ownership of, say, their left arm, and although they can see it hanging there, they make no use of it and may even insist that it must belong to someone else. They have lost the ability to conceive of their left arm. That bit of their body image is gone.

As a kind of perception, we should try to understand the body image in terms of the ideas discussed in chapter twelve. In its unconscious form, the body image can be seen as a representation or model of the body in terms of brain circuitry, nerve-impulse patterns, and so on. And then, never far off, when we become more or less conscious of our actions, there is an evanescent qualia-model too, another version of parts of the unconscious model perhaps but made of feelings and sensations that come and go. Now if we recall the ultimate purpose of perceptual representations, the body image begins to seem like more than just an image, more than a passive perception. A perceptual model, whether conscious or unconscious, cannot be a passive thing requiring another perceiver making a similarly passive model to perceive *it* and so on and on and on. That is where madness lies—or at least a nasty attack of polyhomunculophobia (as described on page 146).

You remember how this condition can be cured? It is by recognising that at some stage a perception must *arrive*, as it were. It must fuse with the self—become part of the self—the Greater Self and then sometimes also the Evanescent Self. At the level of the body image we can announce these arrivals: It seems clear from the strange pathological examples just referred to that our conscious/unconscious body image is part of our conscious/unconscious mind. It is *part* of our mind, part of our self, not something our mind looks at. So one should perhaps be thinking of the body image as working models like the kind we were discussing in the last chapter and which Kenneth Craik was imagining half a century ago. That is to say we should be thinking of models that have their own workings, and that are agents as well, that can operate on the world.[79]

For the unconscious part of the body image, we already have some of idea of what the "workings'" will be like: They will presumably be explicable in terms of nerve impulses, synapses, and so on. How such a representation could have effects is not baffling. Even a static image or "picture" may be able, in a sense, to *do* things—if it is an instruction, if it is wired appropriately, and if the whole thing is supplied with energy. An aeroplane can fly on a course set out on a map of the territory built into its control systems. A scroll of paper punched with holes can play a tune on the pianola. A floppy disk may give instructions to a computer's printer, or control a dextrous robot on an assembly line . . . Similarly, our genes are "agents." And so there seems to be nothing deeply problematical about likening the unconscious part of our body image to a piece of controlling software for our actions: It helps steer us in the things we do "without thinking," and is being updated on details all the time. And if we don't yet know exactly how such software is stored in the brain—what corresponds to the holes in the pianola paper or to the tiny magnetic blips on a floppy disk—we are not exactly baffled. We have some good ideas about what it might be, in terms of persistent alterations of neurons and their patterns of strong connections, even on how memories might be written and read (p. 30).

Of course, the deep problem is with the conscious part of the body image. In the first place, we do not yet understand its physical basis much beyond the bald statement that it is brain-made. The conscious body image is part of our Evanescent Self and so is in some sense "made of qualia." And we should perhaps think of it as another model of the sort that Kenneth Craik imagined for unconscious brain activity: also a control device, also with its own workings, also having effects—although "made differently" (to put it mildly). Anyway, its activities can't be explained adequately in terms of tiny, individual pieces of machinery made from atoms joined variously into molecules that cog together within the wonderful clockwork of molecular biology. Rather, its activities occur from a different way of organising the quantum chaos, achieved nonetheless within this clockwork of molecular biology (or so we were imagining in chapter seven) through clever protein molecules in concert.

The *really* odd thing, if you remember, was how far we had to go from current scientific decency in trying to understand Evanescent Self. It cannot be that these arrangements of feelings and sensations are wholly controlled by the neuro-brain, although they are surely made by the brain somehow and largely controlled by its neural activity. But if feelings and sensations were *just* reflections or shadows of what the neuro-brain was doing, there would be no point to them and so no explanation for their evolution.

So finally, returning to the opening theme of the book, perhaps what seems to be the case here from our inner view of it, *is* the case. Evanescent Self, that changing arrangement of qualia, is by and large self-controlling. It can to some extent rearrange itself. Its body image is the most spectacular part of this self, the part that is perhaps most like an angel. I can hardly believe it, but it seems to be the case: Our conscious body image can move itself—and it is so wired up that when it does so, the real body moves.

C h a p t e r *14*

# A strange device

*all made of passions, and all made of wishes*

*I* use the term "device" to describe our conscious mind because I think consciousness evolved, and because natural selection very much goes in for "clever devices," "technologies" of various sorts. So if the rest of evolution is anything to go by, that's the sort of thing our conscious mind will turn out to be. Evolution does not break any laws of Nature that we know of, but it is a great Contriver, able to use physical and chemical effects in ways that may often *look* like magic.

We have some idea of what the conscious part of our mind is *for*—the first question one might ask about any device one doesn't quite understand. The really tricky question for our Evanescent Self is the next one: "How does it work?" Here we suspected that there are some deep principles of physics and chemistry involved which we have yet to get the hang of—that current science is not up to providing us with an answer to this question. Certainly it still *looks* like magic. But maybe this is being too pessimistic. Very often one can understand reasonably well how a device works without having a deep understanding of the science that may be involved. We can think as engineers, seeing how familiar physical effects are used without necessarily having a deep understanding of how these effects arise in Nature. The electrician who recently rewired our house kept describing electricity as "juice" and warned us that it could "belt you one" if you weren't careful. He was a very good electrician. He didn't need deep quantum physics to do his job. Understanding electricity as an aggressive kind of juice was good enough for his purposes.

Let's think first about a familiar clever device—an electric light bulb. The key physical effect on which a light bulb depends, and which no doubt my electrician friend would have understood well enough, is that squeezing a relatively huge amount of electric juice through a very thin wire will make the wire brilliantly white hot. This might have seemed a pretty crazy idea for making a reliable source of light. Most metals at that temperature melt, and if they don't, they probably burn up more or less rapidly, and anyway such hot thin wires would be very delicate. How are they going to be supported for goodness sake? All sorts of objections come to mind. So there is a whole clutch of auxiliary inventions or ideas needed to get this seemingly crazy idea to work: in the choice of tungsten so that the white-hot wire won't melt; in the need to support the wire and have a sealed transparent envelope to let the light through but keep the air out so the wire won't burn; and in the need to have electrical connectors and insulators.

Now think about another device: Greater Self. We can see this as an enormous computer and the main control device of our behaviour. I think we can identify the key idea here—what might have been written in the first paragraph of Nature's original patent application. It is an idea for sending signals that I never would have thought of: the idea outlined in chapter two of sending a signal by means of a self-propagating change in the voltage across the membrane which separates the inside from the outside of a cell. (Would *you* have thought of that?) The now-unfolding "trade secrets" lie in the details of the membrane's construction—with its ion pores, its pumps, its fuel-supply systems . . .; and in the design of synapses to get long-range signaling to work and perhaps to provide hardware for memory too. And then there are the "handmaiden" roles of glial cells, and other "enabling technology" such as a skull, a circulation . . . All of this is highly complex and sophisticated, and only partly understood—but not baffling, because we think we understand enough of chemistry, cell science, and computing to provide the terms of an explanation.

What then of this conscious device, Evanescent Self? The key idea here is to base the action of the device on conscious phenomena—qualia. Weaving a control device based on feelings and sensations might make lighting with white-hot tungsten look easy. But before we shout "How on Earth!" and give up, let me remind you that it looks very much as though Nature put together such a machine through the process of evolution, as she put together flying machines and other devices that we have been able to copy, or even sometimes excel. Of course, in many respects Nature is still ahead of us. Yet Nature, evolving Nature, the process of evolution through natural selection, cannot be said to "know" any science at all. So I would expect that *in principle* we should be able to do something similar in the matter of making—and making use of—qualia. Here, of course, Nature is way ahead of us, but we should not retire baffled, particularly because there is one aspect of the problem that is not in the least baffling.

Nothing could be more straightforward than accounting for behaviour in terms of things we all have immediate knowledge of. Someone may say: "I was hungry," or "I thought the children would enjoy the game," or "I felt uneasy about Hubert's idea"—and be understood through the experiences we all share—of hunger, fun, or scepticism.

And if we want to be more analytical, we might notice (as in Chapter twelve) that qualia come in a variety of kinds. We might think first of essentially perceptual qualia, the sensations that make up our "multimedia picture" of the world. These are the more passive sensations such as the sensation of a colour, or the feel of a surface, or a sense of distance, or a sense of space, or a sense of motion. They are more about knowing than doing. Of course, action is the ultimate object; it is just that in our complex phenomenal world we are having a good look before we leap.

Yet some perceptual sensations have volitional aspects too. They push us more immediately. They may be nice or nasty in and of themselves—smells, for example. A nasty smell can be very immediately persuasive. But mostly, for humans anyway, it is higher-level combinations of perceptual qualia, actual perceptions, that are more commonly the aversive or attractive items that push us into action one way or another. The object we consciously see—that is, the object in our conscious representation of the world—may be ugly or beautiful, anodyne or terrifying; but the elements of colour and line and movement that make it up are usually none of these things by themselves.

The complicated qualia of thought that come between knowing and doing are a mix of visions and urges: wanting to know, wanting to imagine, wanting to understand; feeling enthusiastic, or bored, or annoyed, or uneasy, or certain. They lead to the higher-order passions and wishes, the mental pains and intellectual pleasures of conscious thought that nudge us, urge us, persuade us, and sometimes compel us.

One can perhaps see a sort of mechanism in all this filled with "components" that are different sorts of qualia—almost like little wheels of understanding, levers of action. We often use such concrete analogies to describe our feelings. Getting more and more frustrated and angry and losing one's temper is like a pot filling up and overflowing, or being heated for too long and boiling over. Or maybe pain is like a force, an almost irresistible force *in extremis*, pushing us. But such things are still only poetic analogies, no use yet in helping us design a real mechanism that might feel angry, to mimic Nature's mechanisms which can make such feelings. The whole way of talking—the causal principles behind, say, the action of pain as one of the elements that contributes to our behaviour—seem so different from the way we usually talk in science as we discussed in the opening chapters of the book. We might say that pain is an experience which is usually *nasty* and add that if it intensifies, it may sometimes add additional notes and become more nasty in more ways—no longer, perhaps, just the old dull ache but some sharper twinges in there too. Pain can be complicated, as well as being a very effective persuader indeed. But its complexity and its persuasive qualities are not well described in terms of arrangements and motions of particles with forces between them. There is nothing nice or nasty about what atoms and molecules do—or cells, or circuits . . .

Before we are tempted to abandon this engineering approach to the problem of consciousness, let us remember the conclusions of chapters five and six. Pain *is* part of the mechanism of behaviour. Such feelings must have real physical effects, not just vicariously or by proxy. They cannot just be another way of talking. The *painfulness* of pain must have a point: the distress of real hunger, the enjoyment of eating, and so on. It cannot be, as some would have us believe, that it is just the "neural correlates" of such feelings that actually do the trick. If neural correlates could do it all, then the actual sensation would be pointless in evolutionary terms.

It can't be like that *just because* such an account of behaviour would be, effectively, qualia-free. That must be wrong. If Nature's engineer, evolution through natural selection, included qualia as part of the mechanism of behaviour, then qualia must be a part of our explanation for behaviour too.

I am reminded of camping with my family for the first time in France and putting up our new tent—rather efficiently I thought, until my small son pointed to several odd poles and a whole bag of special looking tent-pegs left over. It was a bad sign.

No doubt much of the problem of consciousness is at the "trade secrets" level. But as suggested in Part II of this book, there may be a deeper level of difficulty too: that our physics and chemistry are simply not up to it. They are not rich enough.

Let me expand on this a little by giving you an example of another set of ideas which was supposed to explain everything, but which turned out not to be rich enough: the atomism of ancient Greece (ca. 450 B.C.). According to this idea, the world consists only of Atoms in Motion in Space. The persistence of things comes from the atoms being immutable; the variety and variability of things comes from the atoms being of various shapes and sizes, and being able, through their motions, to hook together in different ways.

Might we describe this as a simplified version of our physics and chemistry? Indeed, and it is brilliant. But it is too simple to work. Our current physics and chemistry is far richer, and has to be in order to explain the facts of astronomy which need gravitational fields or something equivalent as part of the story; or of metallurgy, which needs quantum theory to explain even the most basic fact about metals—that they are strong; or of cell biology, where the atoms and molecules have to be far more varied in their properties than simply having different shapes and sizes, and where again forces between atoms and molecules, different kinds of forces, have to be considered.[80] But we should not be too smug. Maybe *our* physics and chemistry and biology are not rich enough yet to cope with feelings and emotions.

Chapter *15*

# The fabrics of the world

*"thus we may see," quoth he, "how the world wags"*

You know how it is driving along in a motor car. The police can discover just how fast you are going by timing your journey between two points, by bouncing radio beams off your car—they have all kinds of ways. And maybe you get a ticket through the mail two weeks later saying that at such and such a time at such and such a place you were going at such and such speed and it was too fast.

Now you would think it would be possible to get both the *speed* and the *position* of any moving object at the same time, and as accurately as you liked provided your equipment was up to it. But this is not absolutely true. It is nearly true for objects the size of motor cars, so nearly true that any discrepancies are way below the limits of observation. But in principle it is not true for an object of any size, and it has been *shown* not to be true for very small objects like atoms. And it has been shown to be completely untrue for extremely small objects such as electrons—for example the ones that make up the beam that scans the back of a television screen and paints the pictures. These electrons all fly from the back of the television tube to the front. But don't enquire too closely how fast they are going and where they are. Abandon all hope of a moment-by-moment account of their journey. It is not possible to know exactly both the position and the speed of an electron at the same time. It *is* possible to know one or the other as accurately as your instruments will allow. But the more you know the one, the less you can know the other. No matter how good your instruments, this will always be true. Weird, isn't it?

No, sorry, *devastating*. Although for everyday objects the effects are utterly trivial; like lovers in a quarrel, we may feel that what is so upsetting is the *principle* of the thing! How *could* matter be *like* that! Things that are in motion must surely *have* a position and speed at any instant, we may be inclined to shout, even if nobody happens to know what they are! I mean for goodness sake!!

Who says? The best theory of matter that we have says, that's who: the quantum theory developed during the first three decades of the 20th century. Among many other things, this theory has explained the basic facts of chemistry; how there can be all these different kinds of atoms, different chemical elements; and how such atoms stick together in complicated ways to make liquids and solids, skin and bone and brains, and apples and people.

What does this have to do with consciousness? It is often said that bringing in quantum theory to try and understand consciousness is just crassly lumping together two difficult areas: trying to use one set of phenomena that nobody really understands to illuminate another set of phenomena that nobody understands at all. Put that way, it doesn't sound very promising. But who knows? We might end up with just one thing nobody understands, rather than two. That would be OK. But that is not my reason for finding quantum theories interesting. My main reason is more straightforward. It is that the problem of consciousness, the question of the nature of the Evanescent Self, is most of all about how a piece of *matter* in your skull can develop feelings and sensations which have physical effects. To have a go at this problem, we should surely be prepared to deploy the deepest theory of matter there is.

There is another more general reason to be interested in the quantum theory of matter in this context. Quantum theory is a fascinating example of a crazy-seeming set of ideas that works very well—and maybe consciousness will need a crazy-seeming theory of some sort, whether this is a quantum theory or not. You see, what is so devastating about quantum theory—in, for example, our failure to determine simultaneously the position and speed of very small objects—is that it is not a failure of instruments, but a failure in ourselves—a failure *about* ourselves. You know these models of the world we were talking about earlier, these models in our heads that are partly neuro- and partly phenomenal? They are *ourselves* more or less, but I'm sorry to say they are wrong. We should have expected it really. All models have their limitations. But ours are wrong in the most startlingly fundamental respects. And of course the models in our heads give us no intuition of this failure. They would be the last to let on.

This inability to precisely and simultaneously determine the position and speed of a very small object is an example of what is called the uncertainty principle or *the principle of indeterminacy*. It turns out that we cannot know everything that our built-in models of the world lead us to expect we should be able to know.

Here is another example. Electrons are fundamental constituents of all the matter on the Earth's surface. Atoms have electrons in them. A carbon atom for example, has six electrons which were at one time thought to be circling around a nucleus like tiny planets circling around an even tinier sun. But an electron is not really an object as we might imagine it—like a grain of sand, for instance. The quantum theory has strong rules about how electrons can be arranged in atoms. But to be in definite places is not one of them. We can only say that an electron in an atom might be found here or it might be found there . . . One can assess probabilities. In principle one can say *exactly what the chances are* that the electron will be found within some specified space near the nucleus. But one can never say, even in principle, *exactly where the electron is*. There is still a combination of "chance" and "necessity," but the chance element is no longer simply a matter of being unable to find and follow all the myriad causes of events. It is not just a practical inability to find out what's what—not chance by default, but a more deeply built-in "chance-pure-and-simple."

If you have ever played with magnets, you will have had a vivid impression of the reality of a magnetic field. It seems to be some kind of tension, a field of influence that exists in the space surrounding a magnet, particularly with the ends of two magnets that repel each other: There is the impression that a magical substance lies between them. And then there are the electrostatic forces, electrostatic fields that can mediate attractions and repulsions, this time between electrically charged objects.

In the 19th century, Michael Faraday showed exactly how these two kinds of force are connected, how changing electric fields of force will produce changing magnetic fields and *vice versa*. He described this in terms of a single, more complicated *electromagnetic field*. These studies led Faraday to invent the transformer and the dynamo, which were key devices in the foundation of the modern electricity industry. Then Faraday's great admirer, James Clerk Maxwell, made one of the greatest discoveries of all time when he realised that light consists of waves in an electromagnetic field. By the end of the 19th century, the question had been settled conclusively: A beam of light is a train of *waves*, electromagnetic waves.

At the start of the 20th century, Max Planck and then Albert Einstein showed that energy is, in fact, like matter insofar as it comes in tiny little packets somewhat analogous to atoms. Light is produced in such packets or "quanta," and it is absorbed in "quanta" too: *photons* they are called. Sorry, did I say that a beam of light is a train of waves? No, no, it consists of photons—a beam of light is a stream of *particles*.

Only joking. Light is, er, both, that is to say neither. It's a sad thing, but our brains are not made to imagine vividly what light is. They are not made to visualise its nature. We can only say that in some ways it's like this, in other ways it's like that. Let's just say that light is *a subtle form of matter*.

Now the plot has thickened. The electromagnetic field is implicated in the forces which allow atoms to combine to make the ordinary materials around us, including molecules such as DNA and proteins; and then for such molecules to be parts of cells; and for cells to come together to make—well, *brains* among other things. Mostly your brain belongs to what is called "the electromagnetic domain." This is a domain of discourse: Such forces predominate in discussions of what the brain is and how it works. Yes the brain being a fair size, is significantly affected by the Earth's gravitational field—it has some weight, that is—but this is not usually reckoned to be a particularly crucial property of brains that will give us blinding insights into how it works. Nor are subatomic nuclear forces seriously implicated in the brain's activities, although they exist in the atoms from which brains, saucepans, and pairs of shoes are ultimately made. We already have major insights about the brain's operation in terms of electromagnetic phenomena: It is a reasonable guess, although only a guess, that the brain's consciousness is also "in the electromagnetic domain."[81]

It is not only that photons are the quanta of light. They have a more general significance: They are "the quanta of the electromagnetic field." You can't get away from them. Short-lived photons are the "carriers" of such forces. They are called "virtual photons." They are permitted, indeed required, by the principle of indeterminacy. They do not necessarily move in straight lines and may cover only tiny distances. They flicker into and out of existence, pushing and pulling the atoms and molecules that produce them and absorb them. What we call empty space is a maelstrom of such "particles"; a chaos from which the fabrics of the world are woven.

I am among those who like to imagine our consciousness as being made of one of these fabrics, and to say that consciousness, like light, is "a subtle form of matter."[82] It is not made out of atoms and molecules but is derived from the same substratum—that the organized qualia of consciousness are to be identified with particular organisations of this maelstrom; so that it is somewhere here, this stuff of consciousness, this fabric that the brain makes and manipulates in making its phenomenal world. This is very different in most respects from, say the white-hot tungsten in an electric light bulb: but it is similar in that it takes much contrivance to produce it and to handle it—a material that can *be* our feelings when properly organised, and has its own properties, its own rules and licences, its own kinds of cause and effect.

The STUFF of consciousness! FABRIC! Do I really mean that?

Of course. The words are chosen for their informality: but you know there is nothing ordinary about *ordinary* stuff. It has turned out to be as problematical as anything else. The trouble started in 1905 with Albert Einstein again, and his strange idea that matter and energy are equivalent. Matter and energy, stone and light, have equal claims to stuffhood it has turned out. So to say that consciousness and its qualia are a kind of stuff is to be rather conventional. It is much the same as saying that qualia are physical effects. What else could they possible be? Incomprehensible, did I hear someone say? Well maybe, but only in the trivial sense of not clicking neatly into a world model built into us at birth.

As a chemist I am perhaps more wedded than most to molecular descriptions and explanations, and see in molecular biology the best explanation we have of LIFE. Of course, more than chemical ideas are needed to explain life—from the theory of evolution in particular. But it is usually supposed that the chemist's kit of ideas provides a structural foundation *beneath* which, in biology, one seldom has to look. This kit of ideas contains *molecules*, which we can usually think of as objects consisting of atoms joined up in particular configurations, and which can be displayed as models consisting of differently coloured balls joined together with little tightly wound springs. Molecules moving, spinning, vibrating; molecules of known sizes and shapes exerting calculable forces on each other, molecules colliding with each other, controlling each other, locking into each other to make the fabric of living matter. We can understand these things remarkably well using the same ideas we use to understand the simpler fabrics of gases, liquids, and solids. So let's think about those for a moment.

A *gas*, such as the air we breath, is a form of molecular matter with the simplest fabric and the most easily deduced properties. The pressure, the temperature, and the volume of a gas can be directly related to the random perpetual motion of its molecules, colliding with each other, bouncing back off the walls of the vessel holding the gas—a kind of fantastic game of billiards with no players. In *liquids*, the molecules are usually bigger, move more slowly, and keep closer to each other; held through weak attractive forces, but able to slither over each other like snakes in a bag. Solids have their atoms more rigidly held in place, and they are more regimented—although regimented in many different ways in different kinds of solids. The various structures of solids are often still quite well represented using little balls for atoms and lots of little springs to show how forces are holding them together in their particular way, while at the same time allowing the molecules to vibrate—for example by a spring between two atoms stretching and contracting, or alternatively vibrating by bending back and forth one way and then the other. Such "vibration modes" store energy, more and more at higher temperatures where the vibrations become increasingly vigorous.

Our toy models work well in explaining much of this. But suddenly the luck runs out. At different levels on different scales, the ball-and-spring models of matter let us down. To explain some properties one has to go back to a deeper level of explanation—the quantum level—and rebuild ideas from there, modifying or even abandoning our "just pretend" ball-and-spring models. For example, the idea of balls and springs moving and vibrating more and more vigorously as the temperature rises gives us a vivid image of what heat *is*, but there is no inkling of an important proviso: these ways of holding energy are *quantised*, to use the jargon. That is to say a bond between two particular atoms can only vibrate with a limited number of discrete levels of vigour—the vibrations cannot just gradually get more and more vigorous as the temperature rises as you would expect from a simple ball-and-spring picture. This is one of the consequences of the quantisation of energy. Another is that energy is always exchanged in packets, the sizes of which are rigidly controlled by local circumstances.

There are several familiar or unfamiliar effects that can only be understood by taking into account the quantisation of energy.

For example, you will not be able to understand the changing colours of coals as they get hotter and hotter without the quantum idea, nor the weird behaviour of helium at exceedingly low temperatures, nor the conduction of electricity through wires—even at ordinary temperatures.

All this is not to say that we cannot use familiar ideas in making new models and analogies in trying to explain to ourselves what the world is like. To return to an earlier example, there is nothing problematic about *force*. It is a familiar idea. We understand it directly when we push something to get it to move. And then *space* is a familiar idea too, very much one of those notions that was bundled in with our brain at birth. But the combined idea, the idea of a force acting across empty space is, on the face of it, plain crazy. Yet this crazy idea works wonderfully well as a way of explaining in detail how apples fall and planets move, and why the tide goes out.

And then again, a *wave* in the sea is a familiar thing and so is a grain of sand on the beach: a *particle*. But for something to behave both like a wave in some respects and like a particle in other respects—well that's bizarre too. Still, this "combination model" also has its uses in helping to explain light and how matter behaves on a very small scale.

Now consciousness is no doubt one of the places where the luck has run out, and the good old balls and springs are going to let us down again. In trying to understand consciousness, we are not, I think, particularly hampered by some kind of logical paradox inherent in a brain trying to understand itself. We can think of someone else's brain if we want to avoid these worries, and be helped indeed by our self knowledge. No, I believe that the problems here are the same kind we have in understanding fundamental science in general. It has a lot to do with finding suitable models. In the first place, not only balls and springs, but ideas such as *substance* that were built into us at birth will, quite likely, not be directly applicable. But there is still every reason to hope that suitable "combination models," however crazy, can be cooked up as food for our imagination.

Yet perhaps you feel that the proposition that qualia are going to be part of physics and chemistry in the future has a bitter taste, raising unpalatable ideas. What is to stop us from saying that all matter is conscious? And where would that ever get us?

Perhaps there are flickers of sensation and feeling in all ordinary physical and chemical processes. Who can say? It is not unreasonable. We do know of a class of physical objects—brains—that produces them hugely.

And is it not remarkable that our brains are so chemically unremarkable? I mean they are made of essentially the same materials as our other organs—indeed, the same basic materials as all life on Earth: proteins, lipids, carbohydrates . . .; phosphate, sodium, potassium . . . These basics were "chosen" by microscopic creatures two or more billion years ago, *long* before the simplest neuron was invented through the process of evolution. Having been "chosen" all that time ago, these materials became virtually fixed, and were to be fixed into the general design of a cell—"the eukaryotic cell"—that was to be the basis of all multi-cell organisms. Plants such as grass and trees, fungi such as mushrooms; animals such as you and I—all can be described as vast communities of eukaryotic cells. These cells are made of a great variety of types for sure, but they are the same basic design. The cells in our brains are no exception. How extraordinary that somehow a suitably organised kit of parts, put together and settled on by micro-organisms, should form the basis of a conscious control computer. It seems as though it hardly matters what stuff a conscious control device is made of so long as its parts are suitably organised.

So what might "suitably organised" mean—not only to be a first class computer, but to be able to generate feelings as well? If that second aspect is a question of organising the maelstrom that is part of every iota of matter and space around us, then, well, perhaps we come back to protein molecules in organisms as the masters of organising matter on a very small scale; that feeling and sensations arise through the complex but concerted activities of large numbers of suitably designed protein molecules; that one quale differs from another by being a different configuration of the maelstrom produced by a different activity of different sets of proteins, each set having evolved, for a somewhat or wholly different purpose; each set having been gradually modified to make a nastier feeling, or a more delicious sensation, and so on; and that there is thus a kind of evolutionary tree of feelings and sensations to be discovered. As remarked in chapter seven, proteins are pieces of submicroscopic machinery which are relatively easily crafted by natural selection and produced by the billion trillions as a matter of course. Living things run on proteins. Well, like all molecules they are chock full of wobbling electric charges and must be spilling out virtual photons like nobody's business—perhaps sometimes in synchrony, in harmony, in dissonance . . .

So the explanation for the invisibility of the machinery that generates qualia, with indications nevertheless of *some* localisation (limbic regions for coercive feelings, V4 for colour, and so on) would then be that these qualagens, these "orchestras" of molecules, are often distributed over fairly extensive regions of brain. The instruments will seem to be ordinary proteins—if you can call any protein ordinary—perhaps doing other jobs *as well*. But what is special is the (qualagenic) organisation of their activities.

Hence, although there may indeed be "flickers of sensation" in molecular processes, generally these would not normally have any effects. If they are not organised, then a host of little sensations and urges this way and that will normally be lost in their mutual contradictions with no large-scale net effects. Think of the glass on the table, which stays motionless although bombarded on all sides by molecules of air: If those molecules *were* organised, as in a tornado (tending to swirl the same way) or as in a furnace (all going very fast) they would smash the glass to pieces or melt it at once.

Perhaps then tiny, random, unanchored scraps of feeling, sensation, volition coming and going in the maelstrom of space became coordinated by what were, at first, chance arrangements of the activities of protein molecules, but which were then established and built on by natural selection . . . And that it was thus that the Great Engineer of evolution hit on an auxiliary-control device of a novel kind.

# Recapitulation

$T$he book was originally intended to be about whether or not we have what is called "free will," and whether science could accommodate such an idea. The opening chapter duly kicked off with this topic. But the discussion soon moved to a related and, it turned out, more approachable question of what real role (if any) feelings and emotions have in human behaviour. In the world according to current science, things are partly determined by laws of Nature and partly by chance circumstance. It is then hard to see how those powerful subjective qualities of, say, pain and pleasure can be brought into the story. I then went on to compare Shakespeare's account of human behaviour—which is largely given in terms of feelings and emotions—to that of neuroscience, framed as it is in terms of nerve impulses and so on. Each account is brilliant, and each rings true. But they are so different. How can they sit together?

The second and third chapters contained a quick rundown of basic brain anatomy, but we also saw how the old idea that the mind and the body inhabit different realms of existence is at last being exorcised from science. And yet we are still left with a view of ourselves as partly conscious and partly unconscious—an Evanescent Self as I called it: perceiving, conscious, wilful; and a Greater Self underlying this: unconscious, more permanent, more intelligent, and more understandable in terms of current science. By the third chapter we were right into feelings, sensations, emotions . . ., what philosophers call *qualia.* The brain makes these for sure. Yet how will they ever be explained in terms of molecules, and cells, and circuitry? There is no escaping them. Qualia are crucial features of consciousness. Indeed I suggested that their presence was the defining feature of consciousness. Perceiving or thinking or acting can sometimes happen unconsciously, and these activities are always *more* or *less* unconscious—which, according to the terminology I am using, means simply *more* or *less* associated with qualia.

This was followed by the three sceptical essays of Part II, the first of these being on the limitations of our present understanding of Nature. It is a common idea that feelings and sensations somehow don't belong in The Physical World, but it is much more interesting to notice instead that somehow they must—because the brain makes them and *it* belongs to The Physical World. The way ahead is to recognise that The Physical World of today's science is a transient edifice. We should expect that what is meant by "The Physical World" will change in such a way that, for example, a pin (the object) and a pinprick (the feeling) will come to be describable in the same language.

Next was an essay describing, briefly, Darwin's idea of evolution through natural selection; and setting out an argument, based on this, that qualia must indeed be part of a physical world *of some sort*. Qualia such as pain or hunger or lust are so well adapted to survival and reproduction, that it looks as though they are products of evolution. But if qualia evolved, they must have both physical causes and physical effects, and so belong to a physical world. I call this "the bomb in the foundations of science"—because the physical world, as we understand it today, has no place for qualia.

The third essay of the trio "Straight talk, double talk, fast talk," attempts to block up bolt-holes. It may seem *so* odd to try to put "subjective" things like pain into the same box as "objective" things like forces between atoms and molecules that we may feel the need to talk our way out of it at all costs—by saying perhaps that emotional and molecular explanations are just different aspects of the same thing, in the kind of way that gas pressure can be seen as being *the same thing* as lots of collisions of molecules. But this is a dubious comparison. Unlike the state of a gas, a state of consciousness has no transparent connection with underlying processes, molecular or otherwise.

Then in part III we were back in the brain, rummaging about to see if we could find the qualia-making machinery. We gave these pieces of machinery a name, "qualagens," without knowing what they were or how they worked. Our preliminary conclusion was that qualia seem to arise from vast numbers of things happening on the scale of molecules, maybe in protein molecules. A particular feeling or sensation arises when a particular group of protein molecules over a sizeable region of the brain oscillates in a particular way, "like swathes of musical instruments contributing to a large-scale effect." It was all becoming very speculative, so we moved onto the firmer ground of the control of conscious states—onto how we wake up, the extraordinary effects of drugs such as cocaine and LSD, and effects of other potions that have been known for thousands of years: All these provide strong indications that the control of consciousness is largely chemical through a variety of molecular key switches. Almost invariably, these switches act on protein molecules in the first place.

The arrow diagrams of chapter nine were intended to express relations between unconscious and conscious aspects of the brain's workings that our evanescent conscious states are largely controlled by unconscious brain activities; that conscious states nevertheless persist for a finite time, and are effective through signals sent back again to the greater, more permanent unconscious part of us.

Part III ended with two chapters on perception, mainly visual perception about which most is known. It first gave an account of how a pair of apparently perfect images of what we are looking at, images on the retinas of our eyes, is analysed. It explained how those images are taken to pieces, as it were, with such aspects as colour, contrast, movement, and so on being dealt with in somewhat different—sometimes completely different—parts of the brain. It ended with ideas about how these diverse computing streams can be used to create the apparently well-integrated world of our conscious experience.

And so to Part IV on "The Secret Agent" of our conscious minds, starting with chapter twelve on the question of why we are conscious at all. The first answer given was that the world of our consciousness, our phenomenal world, is a representation or model of the real world to which it corresponds in some respects. It is a good idea for a sophisticated control system to have such a model. But then the unconscious neural machinery in the brain has such a model too. So the question became: Why have this other one? What are the particular virtues of consciousness—of a model that is in some sense "made of qualia"? The chapter ended with an evolutionary speculation that the first qualia to have appeared would have been pleasures and pains—nice and nasty qualia. For these, the "why" question is directly answered by our personal experiences of such qualia. Perceptual sensations are mostly more neutral, and they evolved that way because they were being used to build a picture of the world, not to be directly persuasive. Then real conscious thinking would have evolved last since it uses both of these kinds of qualia.

Perhaps we could be said to have met The Secret Agent in person in chapter thirteen. Our "body image" is the most intimate part of our model of the world. It is far more than just an image. We saw it as the part of our Self that most immediately acts on the world. Then in chapter fourteen we tried to get some idea of how the Agent might operate, thinking of it in engineering terms as part of an auxiliary control device. And then finally in the last chapter, we extended our earlier discussion on how ideas about the nature of matter have been changing; how the fabrics of the world can now be described as woven from quantum energy; and how perhaps our qualia represent another kind of fabric—the fabric of our conscious minds.

What about free will? Oh, that was left aside—until the Coda, which follows.

# Coda

We return then to our opening theme. Do we have free will? Recall the usual reason for saying no. If our actions all arise out of necessity, there is no free will. If they happen by chance, there is no free will either—and no combining of one with the other will make any difference. But what if our choices are affected by how we feel? Are we not at least *nudged* by feelings most of the time?

But then you might say that, yes, we did what we felt like doing when we stopped for that ice cream and had a cup of coffee as well. But you might say that what we felt like was determined by our brains operating through chance and/or necessity like everything else in the world, and that the way we felt caused or contributed to our actions by a kind of mental clockwork, one that depended on "mental laws." In that case we would have no free will. For example, we could never avoid giving in to temptation. Maybe I try to deny this by saying that I resist little temptations all day long.

But then you say that I am fooling myself. You say that although I might have decided against having a second ice cream, I was just giving in to another temptation, the temptation to be smug and sensible; that I had the warm feeling of righteousness in my sights and that this happened to have been the stronger temptation in the end—that time anyway. We could bat back and forth like this getting nowhere, but I would have failed to make my point. Yes, it might be that the feelings that go toward a decision are effective, but they are not the true causes of the decision. They may just be intermediaries, the way the brain happens to get things done in certain circumstances, the kind of wires that are being pulled.

What other cards might I have to play? In the last chapter I referred to the quantum theory as the best theory of matter we have, and to the critical role of chance in this theory "chance-pure-and-simple" as I called it. This gives us a picture of the Universe in which the future is not just unpredictable in practice but essentially unpredictable. The future does not exist in any sense except as an enormous, intractable set of possibilities.[83] Fatalism is wrong. So might there be room then for passions and wishes to be a part of what decides what will be?

Well, perhaps. But my card turns out to be a somewhat dog-eared five of diamonds—it has been often used, and it might take the trick, but that would depend on other unknown factors coming right for it. You see, quantum theory has its laws too: *tendencies* for different things to happen are calculable, in principle, to as many decimal places as you could want, even if you never quite know which of several possibilities is going to happen in fact. In quantum mechanics as presently formulated, there is no obvious loophole here, because if a feeling could consistently improve the chances of a set of brain molecule behaving one way rather than another, then it would be affecting the tendency for this behaviour. The tendency would then no longer be calculable: Statistically the molecules would appear to be disobeying quantum mechanics. This might be what happens, but as far as I can see, it is not what quantum mechanics as presently formulated would allow.

I might concede all this—but still insist that a revolution in science will be necessary just to give feelings *any* role. "New physics" will be needed. It looks like it, because today's "old physics," including quantum theory was designed to exclude these very phenomena. Feelings and sensations are not treated in "old physics." But then you might say: "Old physics, new physics. What's the difference? It would all be *physics*, wouldn't it?"

Yes, but let me remind you of the centre of all this—the key argument on which my whole story depends, and which you should throw all your wits at if you wish to defeat the idea that feelings and sensations *as such* have functions: that these and other qualia are not just alternative descriptions of neural events which can be fully understood without actually bringing qualia into the discussion. The key idea is this:

## Qualia evolved

I can't prove it, but it is surely the default position to take at this stage in our enquiries. Qualia are well adapted, and suitably wired up. This strongly suggests that they evolved through natural selection. For science, there is no other engineer in Nature.

We hardly need reminding of how well adapted qualia can be; of the pleasures that persuade us to do strange things, for example to seek out one-time living materials and grind them in our toothed mouths and swallow . . .; or the emotions and sensations that persuade us to go through the elaborate quests and contortions that lead to the reproduction of our kind. And you know about pain and anguish—how they may protect us and warn us . . .

Qualia evolved: so they have effects in the physical world. That is the critical conclusion. Qualia have physical causes and physical effects and yet these are different in kind from the causes and effects in the rest of the physical world. At least so it seems to us in our present state of understanding Nature. They are not understood yet in this way, as part of science, and yet they are part of Nature's thoroughly physical process of evolution. So they ominously point to inadequacies in our understanding of the physico-chemical world.

There's a quick reason you might again give for throwing feelings out of the argument. They are *subjective* while proper science is objective. They are outside science in some more fundamental way. You can't get at the subjective world. You can't measure it, do experiments on it, and so forth. But this line of argument does not make sense in the light of evolution. It does not make sense if qualia evolved. It's the same old argument: If they evolved, they *do* have physical consequences (in brains anyway). And that means they are objective *as well*. They have an objective face, as it were.

The brain being an evolved object, we were naturally thinking of it as a device, a device that makes qualia among other things; as a piece of natural engineering which we don't quite understand. Indeed, you don't see much going on even if you open up the skull and peer in. Indeed, it can be notoriously difficult to know whether someone in a coma, for example, has any consciousness or not. But then unless you are an electronics engineer, you probably don't know exactly how a video recorder works either, even if you can correctly manipulate the external knobs and switches. You might not know that anything was happening, if it were not for the whirring noise and perhaps a red light on the front of the machine. But these are trivial indicators of what is really going on inside that black box as it takes in a tangle of radio waves, teasing out and holding onto a signal from which it will be possible to make noisy moving images of Butch Cassidy throwing his weight about. Opening the box and looking inside isn't worth the risk of electrocution: There will be no sign of Butch Cassidy. A machine like this can be doing things that are "occult" in the old meaning of the word: hidden, not evident from the outside. But I assure you that your recorder is being perfectly "scientific." I can't be so sure about your brain, but I suspect something similar there too. There are physical things going on in your brain, some of which constitute your Evanescent Self.

No doubt there *is* something special about subjectivity. Everyone, every creature that is conscious, has direct access to part of its own being. It experiences its own nature—because, in part, it *is* its experiences. But this is all extra stuff from the point of view of understanding. It is an additional strange kind of direct knowledge, subjective knowledge. It does not, in principle, *lessen* our ability to understand the nature of the Evanescent Self as part of the physical world. It is just that we don't happen yet to have got the hang of how it works in objective physical terms—which the evolutionary argument assures us must be possible, even if a "new physics" may be required to do it.

There is a sense in which we might be said to have free will without supposing that a "new physics" (or something) will come to the rescue and reveal a loophole to escape the tyranny of chance and causality.[84] The tyrant has anyway been loosing his teeth recently. People have been realising that it is only in some—usually rather contrived—situations that consequences can be predicted in detail. It has turned out that in many systems—and the brain is almost certainly one of them—you cannot predict what will happen to them because you cannot know enough about its state at any particular moment, never mind calculate all the effects of tiny disturbances on what will happen next. There are plenty of examples where seemingly trivial differences in initial conditions, or seemingly trivial disturbances along the way, can have enormous consequences, producing sometimes altogether different outcomes. For this reason, for example, it is impossible to make precise long-range weather forecasts.[85]

So at least we have Pragmatic Free Will if our brains are at all like the weather in this respect. In practice, it would be impossible to predict exactly what a brain will think; and if such a brain announces that it exercises a free will, it's not a bad idea just to agree.

Traditionally, free will was associated with consciousness, because consciousness was "the mind" and untrammeled by the laws that constrained mere "matter." Having moved from that position, we should now ask again if there should be any necessary connection between free will and consciousness. Recall that most of what we *do* we do unconsciously—in riding a bike, walking across the room, or whatever. In doing such things, we are assessing complex situations and making vital little decisions about these situations unconsciously all the time, and acting on them. Looking from the outside, it may be hard to know which actions are under conscious control and which are not. Why do we tend to use the term "free will" only for the more conscious phases of our behaviour?

I suppose it is because when we are operating in conscious mode we are to some extent doing what we feel like doing, whether or not these feelings ultimately depend on "chance and necessity." In the broader view of the physical world that we can foresee, within which feeling and sensations have effects, then whichever way we look at it, we can, to some extent, do what we want. That is a kind of freedom I could settle for.

# Notes & References

$N$otes on specific points are arranged in a single numbered sequence. More general "Further Reading" citations relating to chapters follow from page 211. Most books referred to are still in print at the time of writing. Journal references are dominantly to *Nature*, *Science*, or *Scientific American*, which can generally be found in science libraries. References given within a descriptive text may be shortened and with modified punctuation.

## Chapter 1  Doing as we like

1 We will leave quantum theory till chapter 15. Another 20th century development is the mathematical theory of Chaos, described in James Gleick's wonderfully readable story (1988), *Chaos: making a new science*, Heinemann. This theory has put the lid on simple Fatalism.

2 Democritus, who flourished about 420 B.C. said this, according to Jacques Monod in his popular molecular account of biology (1970/1972), *Chance and Necessity*, Collins. Empedocles

(ca. 500–430 B.C.) held much the same view: Bertrand Russell (1946), *A History of Western Philosophy*. Allen & Unwin, chapter 6.

3    This is the general theme of Monod's book which is referred to in note 2.

4    *Othello:* Act III. scene iii. line 164.

5    A. G. Cairns-Smith (1996). *Evolving the Mind: on the nature of matter and the origin of consciousness*, Cambridge. Cambridge University Press, p. 108.

6    Gilbert Ryle (1949). *The Concept of Mind*, London: Hutchinson.

7    L. Deeke, B. Grötzinger & H. H. Kornhuber (1976). "Voluntary finger movements in man: cerebral potentials and theory." *Biological Cybernetics* **23,** 99, H. H. Kornhuber & L. Deeke, eds. (1980). "Motivation, motor and sensory processes in the brain: electrical potentials behaviour and clinical use." *Proceedings of the Fifth International Symposium on Electrical Potentials Related to Motivation. Motor and Sensory Processes in the Brain*. Amsterdam: North Holland. See also Benjamin Libet's work: e.g., B. Libet. E. W. Wright. B. Feinstein & D. K. Pearl (1979). "Subjective referral of the timing of a conscious sensory experience." *Brain* **102,** 193–224.

8    H. Pashler (1993). "Doing two things at the same time." *American Scientist* **81,** 48–55.

9    *Othello:* Act I. scene iii, line 321.


## Chapter 2    The two of me

10    Descartes (1637), *Discourse on Method*.

11    Christof Koch (1997), "Computation and the single neuron." *Nature* **385,** 207–210. Terrence J. Sejnowski (1997), "The year of the dendrite." *Science* **275,** 178–179.

12    Eric Kandel & Robert Hawkins (1992), "The biological basis of learning and individuality." *Scientific American* **267, No. 3,** 78–86.

13    Known as Broca's and Wernicke's areas, respectively.

14    Antonio Damasio (1994), *Descartes' Error: emotion, reason and the human brain*, New York: Avon, chapters 1–4.

### Chapter 3    Qualia en croûte

15    *The Oxford English Dictionary (2nd edition)* lists about a dozen distinct, and no doubt legitimate, meanings for "consciousness." As there is no agreed scientific meaning for the word, we just have to be clear about which (if any) of these we have chosen.

16    Hermann von Helmholtz (1821–1894) was perhaps the pioneer here when he suggested that unconscious processes—unconscious inferences—were an essential part of normal perception. It was not a popular idea when it was first proposed: See Richard L. Gregory (1984), *The Mind in Science: a history of explanations in psychology and physics*, Harmondsworth: Penguin Books. p. 362. Unconscious elements in perception are now well recognised: See, for example: Jason B. Mattingley, Greg Davis & Jon Driver (1997), "Preattentive filling-in of visual surfaces in parietal extinction," *Science* **275,** 671–674.

17    Larry Weiskrantz (1986), *Blindsight*, Oxford: Oxford University Press. Jon H. Kass (1995), "Vision without awareness," *Nature* **373,** 195. Alan Cowey & Petra Stoerig (1995), "Blindsight in monkeys," *Nature* **373,** 247–249.

18    How different these pathways are remains to be seen. Other examples of conscious functions being selectively interfered with are given in this chapter and in chapter 6 and 9.

19    N. F. Dixon (1987), "Subliminal perception," in *The Oxford Companion to the Mind*, ed. R. L. Gregory: Oxford: Oxford University Press, pp. 752–755.

20   David Hume (1740/1978), *A Treatise of Human Nature*, Oxford University Press, [Book I; part iv. section vi]. p. 252.

21   John McCrone's engaging book (1990), *The Ape that Spoke*, Picador, gives an account of the evolution of our thought processes that brings in feelings and emotions all the time. Antonio Damasio sees emotion as an essential part of human reasoning: (1994), *Descartes' Error: emotion, reason and the human brain*. New York: Avon: as does Daniel Goleman (1996), *Emotional Intelligence: why it can matter more than IQ*, London: Bloomsbury Publishing.

22   W. Penfield (1959), "The interpretative cortex," *Science* **129,** 1719–1725.

23   Refers to David Chalmers' remark about "the hard problem" of consciousness—how physical processes in the brain give rise to subjective experience—in (1995), *Scientific American* **273, No. 6,** 62–68.

24   Oliver Sacks (1995), "The case of the colour-blind painter," in *An Anthropologist on Mars*, New York: Knopf Inc., pp. 1–38.

25   As reported by Jennifer Altman in *New Scientist*, 26th August 1989, p. 30: "Scanner reveals brain's colour processor"—on work by Semir Zeki. Richard Frackowiak and others at University College London. Sec C. J. Lueck, *et al.* (1989), "The colour centre in the cerebral cortex of man," *Nature* **340,** 386–389.

26   Gottfried Schlaug. Lutz Jäncke. Yanxiong Huang & Helmuth Steinmetz (1995), "*In vivo* evidence of structural brain asymmetry in musicians," *Science* **267,** 699–701.

27   A vivid historical account of these discoveries is given by Richard Restak (1994) in *The Modular Brain*. Touchstone, chapter 11.

28   A. D. Craig, M. C. Bushnell, E.-T. Zhang & A. Blomqvist (1994), "A thalamic nucleus specific for pain and temperature sensation," *Nature* **372,** 770–773. A number of cortical areas may also "light

up" with pain stimuli, but the *affect* of pain—what makes it really nasty—has been associated with activity in a deep part of the cortex, the anterior cingulate. See: Pierre Rainville. Gary H. Duncan. Donald D. Price, Benoit Carrier & M. Catherine Bushnell (1997), "Pain affect encoded in human anterior cingulate but not somatosensory cortex," *Science* **277,** 968–971: and A. D. Craig. E. M. Reiman. A. Evans & M. C. Bushnell (1996), "Functional imaging of an illusion of pain," *Nature* **384,** 258–260.

29   Antonio Damasio (1997), in "Towards a neuropathology of emotion and mood." *Nature* **386,** 769–770; comments on the paper by Wayne C. Drevets. *et al.* (1997), "Subgenual prefrontal cortex anomalies in mood disorders." *Nature* **386,** 824–827.

30   The comments are taken from a table in the paper by R. G. Heath (1963), "Electrical self-stimulation in the brain of man." *American Journal of Psychiatry* **120,** 571–577. This table with other comments and information is reproduced in my book (1996), *Evolving the Mind.* Cambridge University Press, p. 157.

31   Antonio R. Damasio (1994), *Descartes' Error: emotion, reason and the human brain,* New York: Avon. pp. 71–73. See also Francis Crick (1994), *The Astonishing Hypothesis,* London: Touchstone Books, p. 267.

32   For this rather abstract "wiring diagram," I used information from John Eccles (1989), *Evolution of the Brain: creation of the self,* London: Routledge, pp. 99–100; and R. Nieuwenhuys. J. Voogd & C. van Huijzen (1988), *The Human Central Nervous System: a synopsis and atlas,* Berlin: Springer-Verlag, p. 348.

### Chapter 4   It's a funny old world

33   Quotations to be found in David Brewster (1850). *Memoirs of the Life. Writings and Discoveries of Sir Isaac Newton,* vol. 2, chapter 27,

Edinburgh: Bertrand Russell (1946). A *History of Western Philosophy*. Allen & Unwin, p. 862: Francis Crick (1994). *The Astonishing Hypothesis*, Touchstone Books. p. 31. For a still grander scepticism, see John Casti's (1989), *Paradigms Lost*. William Morrow, chapter 7. "How real is the real world" (not real at all, he says).

## Chapter 5    The light of evolution

34    William James's arguments for the efficacy of feelings were put most forcefully in chapter 5 of his great book (1890/1983), *The Principles of Psychology*, Harvard University Press; but the key idea had probably occurred to him nearly twenty years earlier: Robert J. Richards (1987). *Darwin and the Emergence of Evolutionary Theories of Mind and Behaviour*, Chicago: University of Chicago Press. p. 433. The neurophysiologist Ian Glynn wonders, as I do, why so little attention has been paid to this question in (1993), "The evolution of consciousness: William James's unresolved problem." *Biological Reviews of the Cambridge Philosophical Society*, **68**, 599–616.

35    As well expressed in the title (and the book) by Richard Dawkins (1986), *The Blind Watchmaker*, Harlow: Longman.

36    The complete DNA sequence of the genes in *E. Coli* was published in 1997.

37    What Damasio means by "Descartes' Error."

38    William James (1904/1977), "Does 'Consciousness' exist?" Reprinted in *The Writings of William James*, ed. J. J. McDermott, Chicago: Chicago University Press. pp. 169–183.

39    The equivalence of mass and energy is enough to make the point. See my *Evolving the Mind: on the nature of matter and the origin of consciousness*, Cambridge University Press, pp. 218–230.

## Chapter 6  *Straight talk, double talk, fast talk*

40  It seems to me that this is just altogether too diplomatic. "You say it your way and I'll say it mine—because like tom*ah*to and tom*ay*to they mean the same in the end." Maybe that's not the way the song went, but I don't think that sort of diplomacy will defuse the bomb.

41  As Jeffrey Gray complains in his article (1987), "The mind-brain identity theory as a scientific hypothesis: a second look." in *Mind-waves*. eds. Colin Blakemore and Susan Greenfield. Oxford: Blackwell. pp. 461–483.

42  Michael I. Posner & Marcus E. Raichle (1994), *Images of Mind*, New York: Freeman. pp. 125–129.

43  Alan Cowey (1991), in "Grasping the essentials," *Nature* **349,** 102–103, notes how widespread "covert" (unconscious) awareness is even in normal life and discusses the paper (referred to in my main text) by M. A. Goodale. A. D. Milner, L. S. Jakobson & D. P. Carey (1991), "A neurological dissociation between perceiving objects and grasping them." *Nature* **349,** 154–156, in which conscious awareness, but not covert awareness, was grossly impaired through brain damage. Goodale et al. concluded that, most likely. "the neural substrates for [conscious] visual perception of object qualities such as shape, orientation and size are distinct from those underlying the use of those qualities in the control of manual skills." In a subsequent comment, Benjamin Libet (1991). "Conscious or unconscious?," *Nature* **351,** 194–195, points out that these results do not necessarily imply wholly separate pathways. Libet bases his remarks on experiments in which thalamus stimulation led to covert awareness of a sensation if the stimulation was of short duration and conscious awareness if the same stimulus was maintained for more than about half a second; Libet et al. (1991), "Control of the transition from sensory detection to sensory aware-

ness in man by the duration of a thalamic stimulus," *Brain* **114,** 1731–1757.

44    Richard M. Restak (1994), *The Modular Brain: how new discoveries in neuroscience are answering age-old questions about memory, free will, consciousness and personal identity*, New York: Touchstone Books. Chapter 3 contains accounts of specifically, conscious or voluntary actions being altered by disease while similar unconscious or "involuntary" actions are normal. The useful term "duality of motor function" appears on p. 39.

### Chapter 7    How might brains have feelings?

45    Francis Crick (1994), *The Astonishing Hypothesis: the scientific search for the soul*, London: Touchstone Books. p. 3.

46    On February 28, 1678 (or 1679), Newton wrote to his friend Robert Boyle about his dissatisfaction with his own ideas on how gravity might operate through some kind of ether: "I have so long deferred to send you my thoughts about the physical qualities we speak of that, did I not esteem myself obliged by promise, I think I should be ashamed to send them at all. The truth is, my notions about things of this kind are so indigested that I am not well satisfied myself in them . . ." From H. S. Thayer (1953), *Newton's Philosophy of Nature: selections from his writings*, New York: Hafner. p. 112.

47    When a nerve impulse—an action potential—passes down an axon tube, it momentarily changes the voltage between inside and outside by between a hundredth to a tenth of a volt. This may not sound like very much, but the axon has a tiny diameter and its membrane is only a few millionths of a millimetre thick. Thus, there are huge voltage gradients across these membranes even when the cell is "resting" (ca $10^5$ v/cm)—a voltage gradient that would be impossible to maintain, say, between two pieces of copper in air under or-

dinary conditions (sparks would fly)—and there are huge changes to these gradients as the impulse passes through. Axons have all the appearance of being specialised to send signals fast to distant places. I would not expect them to be just the thing for making qualia as well.

48 See: Barbara A. Barres (1991), "New roles for glia," *Journal of Neuroscience* **11**, 3685–3694: and more recently: Frank W. Pfrieger & Barbara A. Barres (1997), "Synaptic efficiency enhanced by glial cells in vitro," *Science* **277**, 1684–1687. Each of these papers emphasises the "intimate partnership" of glial cells with neuron activity, the second one *in vitro*. The article by P. Morcll & W. T. Norton (1980), "Myelin."*Scientific American* **242, No. 5,** 74–89, is about the system for insulating long axons. See also H. K. Kimelberg & M. D. Norenberg (1989), "Astrocytes," *Scientific American* **260, No. 4,** 44–52. On the idea of glial cell signaling. see J. W. Dani, A. Chernjavsky & S. J. Smith (1992), "Neuronal activity triggers calcium waves in hippocampal astrocyte networks," *Neuron* **8,** 429–440.

49 Francis Crick and Christof Koch suggest that while we can be said to be conscious of neural activity at "higher" levels of the cortex such as V4 and V5, "lower" levels (V1 and perhaps V2) are *not* associated with consciousness (1995), "Are we aware of neural activity in primary visual cortex?" *Nature* **375,** 121–213. Evidence in support has come from Sheng He. Patrick Cavanagh & James Intrilligator (1996), "Attentional resolution and the locus of visual awareness." *Nature* **383,** 334–337, as commented on by C. Koch and R. B. H. Tootell (1996), "Stimulating brain but not mind." *Nature* **383,** 301–302. Antonio Demasio (1996), "Knowing how, knowing where," *Nature* **375,** 106–107, points out that the neural correlates of visual experience must invoke nonvisual brain regions which include "the entire set of structurcs, from the brainstem on up to the cortex, which map the body of the organism cngaged in seeing and being aware of what is being seen." This must be true if one says, as I do, that a state of

consciousness is an organisation of qualia: Then everything from the eyes to the whole brain is involved in the *organisation* of a conscious visual perception. And recalling the analogy with my doorbell (pages 76–79), there will be correlates all along the way. But I think the real question is: How are the qualia themselves made? Where are the qualagens and how do they work? It is quite likely that there are none in V1. It is very possible that there are some in V4 and V5.

50    Michael I. Posner & Marcus E. Raichle (1994), *Images of Mind*, New York: Freeman, p. 96.

51    John Searle sees the distinction between macrodescriptions and microdescriptions as crucial in demystifying the relation of brain to mind: "Nothing is more common in nature than for surface features of phenomena to be caused by and realized in a microstructure, and those are exactly the relations that are exhibited by the relation of mind to brain," he says in (1987), "Minds and brains without programs" in *Mindwaves*, eds. Colin Blakemore and Susan Greenfield. Oxford: Blackwell, on p. 224. See also: John Searle's (1984), *Minds. Brains and Science*, BBC Publications, p. 76. But of course this is only a first step. The next step, which is often taken, is a false one— or so it seems to me. It is in the identification of the "microstructure": It supposes that the microprocesses that underlie qualia are those we already know about and understand—roughly speaking, neuronal "computing activity." On the contrary, I am arguing in this book that (as with my Christmas telephone, which so mystified me as a child (p. 51), there are principles of physics here which are not understood—not yet.

52    Modern techniques of X-ray crystallography, which are now routine, allow us to examine protein molecules in sufficient detail to be able, very often, to see how they work.

53    A particular idea that interests me was proposed by Ian Marshal (1989), "Consciousness and Bose-Einstein condensates." *New Ideas*

in Psychology **7**, 73–83. This depends on the idea that consciousness is a "macroquantum effect"—like laser action or superconductivity, which depend on the tendency for members of a class of entities called bosons to behave in a unified manner. The idea is discussed by Michael Lockwood (1989), *Mind Brain & the Quantum: the compound I*, Blackwood; and by Dana Zohar (1990), *The Quantum Self: human nature and consciousness defined by the new physics*. William Morrow. Such ideas about the physical basis of conscious states being large-scale effects of zillions of microscopic processes are considered by Roger Penrose (1994), *Shadows of the Mind: a search for the missing science of consciousness*. Oxford University Press, chapter 7, which builds also on Stuart Hameroff's idea that structural cell proteins of microtubules are critical to consciousness expounded in his (1987), *Ultimate computing: biomolecular consciousness and nanotechnology*, Amsterdam: North Holland. See also Mari Jibu & Kunio Yasue (1995), *Quantum Brain Dynamics and Consciousness: an introduction*, Amsterdam: John Benjamins. Quantum theories of consciousness are discussed in my (1996), *Evolving the Mind: on the nature of matter and the origin of consciousness*, Cambridge University Press, chapter 9.

### Chapter 8   Switches

54   Bruno Giros, Mohamed Jaber, Sara R. Jones, R. Mark Wightman & Marc G. Caron (1996), "Hyperlocomotion and indifference to cocaine and amphetamine in mice lacking dopamine transporter," *Nature* **379**, 606–612.

55   *As You Like It*, Act IV, scene i, line 16.

56   There are at least 14 different serotonin receptors in the brain: Antonio R. Damasio (1994), *Descartes' Error: emotion, reason and the human brain*. New York: Avon, pp. 76–77.

57   Jaime Diaz (1997), *How drugs Influence Behaviour: a neuro-behavioural approach*, Upper Saddle River, NJ: Prentice Hall.

58    Peptides are much larger molecules than the monoamine neuro-
      transmitters. They are, in effect, small proteins consisting of rela-
      tively few amino acid units joined in a chain. Peptides are common
      as signal molecules, not only in the brain, but for other control sys-
      tems of the body too. The hormone insulin is perhaps the best
      known signal molecule of this type. Since most of the critical ma-
      chinery of life consists of large protein molecules, it is not surprising
      that smaller molecules of similar make up are particularly effective
      at interacting with, and exerting control on, such machinery. In ad-
      dition there are many possible permutations (about $10^{13}$ for even
      just ten amino acids in a string).

          In the brain there are far more *neuropeptides* known than con-
      ventional neurotransmitters. They are produced at axon terminals
      and so are like neurotransmitters in this respect. But their function
      is less certain. The potentially vast number of peptides might
      make them suitable keys for the kind of intricate switching of a va-
      riety of qualagens that might be needed to define a stream of con-
      sciousness. And neuropeptides are at least *associated* with powerful
      qualia: See Alberts *et al.* (1989): *Molecular Biology of the Cell*, 2nd
      edition: Garland Publishing: New York, p. 1094.—with sensations
      of thirst, sex, hunger, pain and so on—whether or not these pep-
      tides are ever actual switches of qualagenic proteins. 'Substance P'
      consists of a string of eleven amino acids. It has long been impli-
      cated in the transmission of pain signals in the spinal cord, but is
      present in the brain too. For some more recent news on this see
      Leslie Iverson (1998) 'Substance P equals pain substance?' *Nature*
      **392**, 334–335.

### Chapter 9    Arrows and desires

59    Francis Crick in (1994), *The Astonishing Hypothesis*, London: Touch-
      stone Books, pp. 67–70, discusses the distinction between "working
      memory," which lasts a few seconds (and is useful for telephone num-

bers) and a much shorter "iconic memory," a fleeting memory of rapidly fading images, which is perhaps the true measure of the span of consciousness and lasts for less than a second. James Glanz (1998), has a report "Magnetic brain imaging traces: a stairway to memory" in *Science* **280,** p. 37, on work by Samuel Williamson and colleagues using a new technique which detects minute variations in a brain's magnetic fields, and through which it was possible to see when and for how long different parts of the cortex responded to a brief visual stimulus. The primary visual cortex V1 at the back of the brain responded first, but the image was "forgotten" again within tenths of a second. The higher processing centres further forward became active late, and they had longer "forgetting times" too: These were "as long as 30 seconds farther downstream." So perhaps the evidence that conscious experience seems to take a second or so to switch on (note 7) is consistent with the indications that V1 activity is wholly unconscious (note 49).

60   K. A. Flowers (1987), "Parkinsonism," in *The Oxford Companion to the Mind.* ed. R. L. Gregory, Oxford University Press. pp. 587–591.

61   From a footnote in an account of the patient Francis D. in: Oliver Sacks (1973), *Awakenings*, London: Duckworth & (1990) New York: Picador, p 63.

### Chapter 11   *Together again*

62   John Horgan (1995), "The waterfall illusion: an odd puzzle yields clues to consciousness," *Scientific American*, **273, No. 1,** 13–14. Illusions can be especially useful for the study of consciousness. Francis Crick (1996). "Visual perceptions: rivalry and consciousness," *Nature* **379,** 485–486, describes two "illusion" papers as being "among the opening salvos of a concerted attack on the baffling problem of consciousness," namely: David A. Leopold & Nikos K. Logothetis (1996), "Activity changes in early visual cortex reflects

monkeys percepts during binocular rivalry," *Nature* **379,** 549–553: and Nikos K. Logothetis & Jeffrey D. Schall (1989), "Neuronal correlates of subjective visual perception," *Science* **245,** 761–763. See also: Erik D. Lumer, Karl I. Friston & Geraint Rees (1998), "Neural correlates of perceptual rivalry in the human brain," *Science* **280,** 1930–1934. An extraordinary "illusion" of pain can be created by running one's fingers over a grating made up of alternating warm and cool bars, which individually are neither painfully hot nor painfully cold, helping to implicate the anterior cingulate cortex with the perception of thermal pain: A. D. Craig, E. M. Reiman. A. Evans & M. C. Bushnell (1996), "Functional imaging of an illusion of pain," *Nature* **384,** 258–260: commented on by Kenneth L. Casey (1996), "Resolving a paradox of pain." *Nature* **384,** 217–218.

63    Roger B. H. Tootell, et al. (1995), "Visual motor aftereffect in human cortical area MT revealed by functional magnetic resonance imaging," *Nature* **375,** 139–141.

64    Richard M. Restak (1994), *The Modular Brain*, New York: Touchstone Books, p. 29; R. H. Hess, C. L. Baker & J. Zihl (1989), "The motion blind patient: low level spatial and temporal filters," *The Journal of Neuroscience* **9,** 1628–1640.

65    Michael I. Posner & Marcus E. Raichle (1994), *Images of Mind*, New York: Freeman, p. 15. Leslie G. Ungerleider (1995), "Functional brain imaging studies of cortical mechanisms for memory," *Science* **270,** 769–775.

66    Yasushi Miyashita (1995), "How the brain creates imagery: projection to primary visual cortex," *Science* **268,** 1719–1720. Alumit Ishai & Dov Sagi (1995), "Common mechanisms of visual imagery and perception," *Science* **268,** 1772–1774. Leslie G. Ungerleider (1995), "Functional brain imaging studies of cortical mechanisms for memory," *Science* **270,** 769–775. John H. R. Maunsell (1995),

"The brain's visual world: representation of visual targets in cerebral cortex," *Science* **270,** 764–769. Alex Martin, Cheri L. Wiggs. Leslie G. Ungerleider & James V. Haxby (1995), "Neural correlates of category-specific knowledge," *Nature* **379,** 649–652.

67   Michael P. Stryker (1989), "Is grandmother an oscillation?" *Nature* **338,** 297–298. Charles M. Gray, Peter König, Andreas K. Engel & Wolf Singer (1989), "Oscillatory responses in cat visual cortex exhibit inter-columnar synchronisation which reflects global stimulus properties," *Nature* **338,** 334–337.

68   Pieter R. Roelfsema, Andreas K. Engel, Peter König & Wolf Singer (1997), "Visuomotor integration is associated with zero time-lag synchronisation among cortical areas," *Nature* **385,** 157–161.

69   S. Ramón y Cajal (1909–1911/1972), *Histologie du Systèm Nerveux de l'Homme et des Vertébrés,* Paris: Maloine, reprinted Madrid: C.S.I.C.

70   David Ferster & Nelson Spruston (1995), "Cracking the neuronal code," *Science* **270,** 756–757.

71   J. J. Hopfield has suggested one such model: (1995), "Pattern recognition computation using action potential timing for stimulus representation." *Nature* **376,** 33–36; commented on by Terrence J. Sejnowski (1995), "Time for a new neural code?" *Nature* **376,** 21–22.

72   Approaches to this combinatorial problem are suggested by M. Abeles (1991), in *Corticonics: neural circuits of the cerebral cortex,* Cambridge, Cambridge University Press: and by Wolf Singer (1995), "Development and plasticity of cortical processing architectures." *Science* **270,** 758–764. See also: Eberhard E. Fetz (1997), "Temporal coding in neural populations?" *Science* **278,** 1901–1902: and Marcia Barinaga (1998), "Listening in on the brain," *Science* **280,** 376–378.

**Chapter 12   Why a phenomenal world?**

73   Francis Crick (1994), *The Astonishing Hypothesis*, London: Touch-stone Books, p. 145.

74   Kenneth J. W. Craik (1943), *The Nature of Explanation*, Cambridge: Cambridge University Press.

75   A finite regression is harmless. See my (1996), *Evolving the Mind*, Cambridge University Press, p. 197. It may be perfectly understand-able if it involves a transduction from one kind of thing to another. Suppose someone asks you to explain how a telephone works. You start by describing how sound wave variations are converted by a microphone into corresponding variations of electric current. . . . Your student angrily interrupts you: "Oh, don't give me that! That's not an explanation. You're just avoiding the issue of how sound can travel so far. These 'variations of electric current' you speak of still have to travel immense distances, and that is just as baffling: You are merely substituting one problem for another." Your student would certainly have a point, but at the same time determined to miss the point.

76   How might one measure anger? One could make a scale. If he starts slamming doors, it's level 3: jumping up and down, level 6: throt-tling people, level 9, and then sometimes he gets really angry. . . . Of course, it is not a very good scale because apart from anything else we know that anger comes in different kinds. If we were to do a de-cent job of measuring them, they should be disentangled as, say, the different forms of energy were in the 18th and 19th centuries.

77   Richard Gregory's piece on "Perception" in his (1987), *The Oxford Companion to the Mind*, Oxford University Press, pp. 598–601, is re-quired reading. On page 601 he says "Looking 'inwards' by intro-spection, we *seem* to know that perceptions are made of sensations, although from physiological and psychological experiments . . . it has to be denied that sensations are the data of perception. The data

are neural *signals* from the transducer senses . . ." Here I am more in-clined to say, with Hamlet: ". . . nay it is: I know not 'seems'—but agree otherwise. Indeed, sensations are not data. But then nor are traveling changes in membrane voltages—nerve impulses—data. These are rather the physical means of carrying and processing the data of our unconscious perceptions. And I would say that sensa-tions are the physical means of carrying the data of our conscious perceptions. Of course, it seems odd to say that sensations are *physi-cal* like this but remember Part II ("the bomb in the basement"): If *sensations evolved* then they are physical things, as much as say, trav-eling changes in membrane voltages even if, for sensations, we can-not yet for the life of us see how. And (chapter 5) it looks as if sen-sations evolved.

## Chapter 13   A working image

78   For example, Ian Howard (1987). "Spatial co-ordination of the senses." in *The Oxford Companion to the Mind*, ed. Richard Gregory, refers on p. 728 to phantom limbs with the remark. "An amputee will attempt to use his phantom limb when doing habitual things."

79   In terms of the diagrams of chapter 9, the C⟶C⟶ C⟶ stream is then a processing stream as the B⟶B⟶B⟶ stream is, and there are also C ⭦ B arrows.

## Chapter 14   A strange device

80   Ancient atomism was not rich enough to explain everything about the material world. For example, Isaac Newton was to add gravita-tional forces operating across space, and he suggested that some such forces would be important in holding the smallest parts of ma-terials together: "Have not the small particles of bodies certain pow-ers, virtues, or forces by which they act at a distance . . ." (1730/1979), *Opticks, (4th edition)*, Quest, 31, London (1730),

reprinted New York (1979), Dover, pp. 375–406. Then Faraday and Maxwell were to see fields to be as real as atoms: M. Faraday (1844), *Matter*, London: Library of the Institution of Electrical Engineers. Planck and then Einstein would upset the whole idea of material substance: Quantum energy became the ultimate stuff whose ultimate grain was action. And then, with Heisenberg, atoms had ceased to be *things* at all: Werner Heisenberg (1959), *Physics and Philosophy: the revolution in modern science*, London: Allen & Unwin, chapter 10. And now we have another quantum physicist, Henry Stapp, with talk of the "feel of events": Henry P. Stapp (1993), *Mind, Matter and Quantum Mechanics*, Berlin: Springer-Verlag.

### Chapter 15    The fabrics of the world

81    But only a guess: Roger Penrose, for example, believes that a future combined theory of gravity and quantum phenomena may be necessary for an understanding of consciousness. See his (1989), *The Emperor's New Mind: concerning computers, minds and the laws of physics*, Oxford University Press: and (1994), *Shadows of the Mind: a search for the missing science of consciousness*, Oxford University Press.

82    The roman poet Lucretius said as much in his work (ca. 55 A.D.), *The Nature of the Universe*, "Book III, Life and Mind." Translated by L. E. Latham, Harmondsworth: Penguin Books, where he says (p. 71) that "Mind and spirit are both composed of matter," mind being a material, ". . . of very fine texture and composed of exceptionally minute particles."

### Coda

83    St. Augustine said much the same 16 centuries ago, as I feel bound to mention in *Evolving the Mind*, pp. 215–218.

84   Marvin Minsky makes a good case against there being a "third alternative" to "cause and chance" in his (1987), *The Society of Mind*. Heinemann. pp. 306–307. We can't bear the idea that our decisions are always the result of forces inside us that we do not understand so we imagine a third alternative: a thing called "freedom of will." Francis Crick, in a postscript to (1994), *The Astonishing Hypothesis*, Touchstone Books, also argues well that free will is an illusion, but then Crick, like Minsky, takes the view that qualia are beyond the pale scientifically, at least at present, and may always be. Ian Stewart and Jack Cohen, in a reluctant but excellent chapter on free will in their (1997), *Figments of Reality*, Cambridge University Press, also believe free will to be an illusion, but more specifically is a quale built into our whole way of thinking.

85   James Gleick's book (1988), *Chaos: making a new science*, Heinemann, starts off with the now-famous "butterfly effect" to illustrate this point. It is not quite correct to say that a butterfly flapping its wings in Brazil might cause a thunderstorm in London a few months later, but it is a brilliant image. And what *does* seem to be true is just as interesting: To predict exactly what the weather will be like at any place on Earth a few months ahead, you would have to know what every butterfly was up to. You would have to take account of every slamming door, every breath, every falling leaf . . . The tiniest swirl of the air cannot help but redirect somewhat bigger breezes already there, which can then alter still bigger air movements . . . So that with an ever-spreading effect, an innocent wing flap will be set to change the world's weather from what it might have been if all else but that wing flap had been the same.

*Further Reading for Chapter 1: Doing as we like*

*Books by scientists on "philosophy of science"*: John D. Barrow (1988), *The World Within the World*, Oxford University Press, on the nature of the world and our understanding of it; his (1998). *Impossibility: the limits of science and the science of limits.* Oxford University Press, is particularly good on the convolutions of the problem of free will. Ian Stewart & Jack Cohen have also written highly readable books about how the world is, but from a more biological perspective: (1994), *The Collapse of Chaos: discovering simplicity in a complex world* Harmondsworth: Penguin, and: (1997). *Figments of Reality: the evolution of the curious mind.* Cambridge University Press. John L. Casti in (1998), *The Cambridge Quintet: a work of scientific speculation*, Addison-Wesley, gives us an imaginary conversation on a fictitious occasion: a dinner party arranged by C. P. Snow in the summer of 1949 with four other thinkers: Alan Turing (computer pioneer). J. B. S. Haldane (biologist). Erwin Schrödinger (physicist) and Ludwig Wittgenstein (philosopher). The subject: whether "thinking machines" could or should be made. Fifty years later the problem of consciousness remains. Perhaps there was something wrong with the direction of these great thinkers' thoughts?

*Books by (mainly) scientifically minded philosophers:* Daniel C. Dennett (1991) in *Consciousness Explained*, Penguin, sees the problem of consciousness as difficult but soluble. There is not much on feelings and sensations here, except to say that they create a tangle that should be cut away: that there is nothing "extra" about consciousness over and above the neural processing, etc. that we already know quite a lot about. I don't agree, but dull would he be of soul who did not find this a riveting book. Dennett's (1996), *Kinds of Minds: towards an understanding of consciousness*, Basic Books, is shorter and also a good read. David J. Chalmers (1996), *The Conscious Mind: in search of a fundamental theory*, New York: Oxford University Press, does not try to talk away the difficulties presented by feelings and sensations. Such things are on centre stage. But Chalmers takes it that physics is essentially complete and so concludes that feelings are not, after all, efficacious. Nicholas Humphrey in (1992), *A History of the Mind: evolution and the birth of consciousness*, Simon and Schuster, takes feelings seriously too ("I feel therefore I am"). This is more psychological than physical, avoids technical terms, and is full of interesting and original examples. Galen Strawson (1994), *Mental Reality*, MIT

Press, has much philosophical fresh air. He does not think that pain, for example, has all that much to do with its outward and visible signs. Michael Lockwood in his (1989), *Mind, Brain & the Quantum: the compound 'I,z'* Blackwood, believes that the best theory of matter we have—quantum theory—may very well have something to say on the question of the relationship between matter and conscious mind. A fascinating book, well written, but quite hard work.

*Further Reading for Chapter 2: The two of me*

Basics anatomies of brains and brain cells can be found for example, in Francis Crick (1994), *The Astonishing Hypothesis: the scientific search for the soul*, London: Touchstone Books, chapter 7 & 8: or my own (1996), *Evolving the Mind*. Cambridge University Press, chapters 3 & 4; or in the special issue on Mind and Brain in (1992), *Scientific American* **267, No. 3.**

*Further Reading for Chapter 3: Qualia en croûte*

I have chosen three books which make important use of case histories—"anecdotal" perhaps, but providing unique and deep insights into the problem of consciousness. Oliver Sacks (1995), *An Anthropologist on Mars*, New York: Knopf Inc: Antonio R. Damasio (1994), *Descartes Error: emotion, reason and the human brain*, New York: Avon: and Richard M. Restak (1994), *The Modular Brain: how new discoveries in neuroscience are answering age-old questions about memory, free will, consciousness and personal identity*, New York. Touchstone Books. I also might mention three texts on imaging methods: Marcia Barinaga (1997). "New imaging methods provide a better view into the brain," *Science* **276**, 1974–1976: Tim Beardsley (1997), "The machinery of thought," *Scientific American* **277, No. 2,** 58–63; and a book I refer to frequently: Michael I. Posner & Marcus E. Raichle (1994), *Images of Mind*, New York: Freeman. Finally, two books on cognitive psychology: The first is a short book of ideas on what this subject is, on its historical origins, and on how consciousness is part of it. George Mandler (1985), *Cognitive Psychology: an essay in cognitive* *science*. London: Lawrence Erlbaum. Then on the distinction between unconscious and conscious "operating modes." and plenty more besides, see Bernard Baars (1988), *A Cognitive Theory of Consciousness*. Cambridge University Press.

*Further Reading for Chapter 4: It's a funny old world*

Try Galen Strawson (1994), *Mental Reality*, Cambridge, Mass: M.I.T. Press, especially the two chapters on "Agnostic Materialism," the second of which ends with the sentence: "Experience is as much a physical phenomenon as electric charge." Strawson, too, is referring to physics as may be, not physics as it now is.

*Further Reading for Chapter 5: The light of evolution*

My chapter title comes from Dobzhansky's famous saying: "Nothing in biology makes sense except in the light of evolution." The light is used, although for quite different purposes from mine, in Michael Gazzaniga's chatty and charming book (1992), *Nature's Mind: the biological roots of thinking, emotions, sexuality, language and intelligence*, BasicBooks, in which both evolutionary and social factors come into the story—as they also both do in Stephen Pinker's blockbuster (1997), entitled simply, *How the Mind Works*, Allen Lane, and in which ideas and stories about the evolution of humanity are used to throw light on the question. John Morgan Allman (1999), *Evolving Brains*, New York: Freeman. The story of the human brain beautifully told in the context of brains in general.

*Toward a Science of Consciousness III: The Third Tuscon Discussions and Debates* (edited by Stuart R. Hameroff, Alfred W. Kaszniak, and David J. Chalmers) Cambridge, MA: MIT Press; in press, due 1999; has contributions on evolution and mind, including my "If qualia evolved . . ."

*Further Reading for Chapter 7: How might brains have feelings?*

Almost any elementary biochemistry textbook, and many chemistry textbooks, explain in more detail about proteins and the way they are made.

Chapter 2 in my *Evolving the Mind* does so, and at a more elementary level my (1985), *Seven Clues to the Origin of Life*. Cambridge University Press, chapter 4. For references to speculations on the physical basis of consciousness, see note 53.

*Further Reading for Chapter 8: Switches*

Mircea Steriade (1996), "Arousal: revisiting the reticular activating system," *Science* **272,** 225–226. Jaime Diaz (1997), *How drugs Influence Behaviour: a neuro-behavioural approach*. Upper Saddle River. NJ: Prentice Hall. Michael I. Posner & Marcus E. Raichle (1994), "Mental Disorders," in *Images of Mind,* New York: Freeman, chapter 9.

*Further Reading for Chapter 9: Arrows and desires*

Moussa B. H. Youdim & Peter Riederer (1997), "Understanding Parkinson's disease," *Scientific American* **276, No. 1,** 38–45; Oliver Sacks (1973) *Awakenings,* London: Duckworth & (1990) New York: Picador.

J. Allan Hobson (1999), *Consciousness*, New York: Freeman: A wonderful book, especially its accounts of the author's use of sleep experience and experiments in formulating "a conceptual model of conscious states" (Chapter seven).

*Further Reading for Chapter 10: Humpty Dumpty*

Francis Crick makes vision his prime example of a conscious mental function (1994), *The Astonishing Hypothesis: the scientific search for the soul,* London: Touchstone Books. A more academic approach can be found in Martin J. Tovée (1996). *An Introduction to the Visual System*. Cambridge: Cambridge University Press. Also see articles by two leading researchers in the field: Margaret Livingstone and David Hubel (1988). "Segregation of Form, Color, Movement, and Depth: Anatomy, Physiology, and Perception," *Science* **240,** 740–749, and also Margaret Livingstone (1988). "Art, illusion and the visual system," *Scientific American* **258,** No. 1, 68–75.

*Further Reading for Chapter 12: Why a phenomenal world*

See P. N. Johnson-Laird (1983), *Mental Models*, Cambridge University Press, and Richard Gregory (1984), *Mind in Science*, London: Penguin Books for accounts of the development of Kenneth Craik's ideas.

Victor S. Johnston (1999) also deals with the title question of this chapter in his lucid *Why We Feel: the science of human emotions*. Reading: Perseus Books.

*Further Reading for Chapter 13: A working image*

Richard M. Restak (1994), *The Modular Brain: how new discoveries in neuroscience are answering age-old questions about memory, free will, consciousness and personal identity*, New York: Touchstone Books, chapter 4: "An existential illness." Another book with case histories is by Israel Rosenfield (1993). *The Strange, Familiar and Forgotten: An Anatomy of Consciousness*. Vintage, very much emphasises the importance of the body image in relation to consciousness and the self. John Searle has a commentary of this book in his own (1997), *The Mystery of Consciousness*, London: Granta Books, chapter 7. Summing up, Searle says: "My conscious experience of my own body as an object in space and time, an experience that is in fact constructed in my brain, is the basic element that runs through all of our conscious experience."

In their brilliant and funny *Phantoms in the Brain*, V. S. Ramachandran and Sandra Blakeslee (1998) give us profound insights, from clinical experience, into the abnormal and normal human self.

*Further Reading for Chapter 15: The fabrics of the world*

The (1989) and (1994) books by Roger Penrose referred to in note 81 include nonmathematical descriptions of quantum theory, but are quite hard going. His (1997), *The Large, the Small and the Human Mind*, Cambridge University Press, is easier. Also recommended: P. C. W. Davies & J. R. Brown (1986), *The Ghost on the Atom*, Cambridge University Press, which is a based on a series of radio interviews with distinguished quantum physicists. The first chapter is a simple introduction to "The strange world of the quantum." Richard Feynman's (1985) *QED: the strange theory of light and matter*, Harmondsworth: Penguin, is written in an easy, friendly style, but gives a profound insight into the nature of scientific theories—especially the ones that sound completely crazy. Arthur Zajong's (1993), *Catching the light: what is light and how do we see it?* Oxford University Press, marches between poets and quantum physicists. Deep popular science. Then there is the classic, relatively late work of Werner Heisenberg (1959), *Physics and Philosophy: the revolution in modern science.* Allen & Unwin.

Three other classics in which scientists ponder the paradoxes of the mind are Arthur Eddington (1928), *The Nature of the Physical World.* Cambridge University Press: Charles Sherrington (1940), *Man on his Nature*, Penguin Books: and Erwin Schrödinger (1958), *Mind and Matter*, reprinted in a 1992 Canto edition of *What is Life?* by Cambridge University Press.

# Index